Linear Dependence

Theory and Computation

Linear Dependence

Theory and Computation

S. N. Afriat*

University of Siena
Siena, Italy

Springer Science+Business Media, LLC

Library of Congress Cataloging-in-Publication Data

Afriat, S. N., 1925-
   Linear dependence : theory and computation / S.N. Afriat.
      p. cm.
   Includes bibliographical references and index.
   ISBN 978-1-4613-6919-6   ISBN 978-1-4615-4273-5 (eBook)
   DOI 10.1007/978-1-4615-4273-5
      1. Linear dependence (Mathematics) I. Title.

QA187 .A37 2000
512'.5--dc21

                                                          00-034933

ISBN 978-1-4613-6919-6

©2000 Springer Science+Business Media New York
Originally published by Kluwer Academic / Plenum Publishers, New York in 2000
Softcover reprint of the hardcover 1st edition 2000

http://www.wkap.nl/

10 9 8 7 6 5 4 3 2 1

A C.I.P. record for this book is available from the Library of Congress

Preface

This work deals with a most basic notion of linear algebra, linear dependence. While providing an accommodation for distinct possible novelties, the purpose is to bring emphasis on approaches to the topic, serving at the elementary level and more broadly. A typical feature is where computational algorithms and theoretical proofs are brought together. Another is respect for symmetry, so that when this has some part in the form of a matter it should be reflected also in the treatment. Then there are issues to do with computational method.

These interests suggested a limited account, to be rounded-out suitably. Hence here we have the early part of linear algebra, where there is the space, base and dimension, and the subspaces, restriction to rational operations, and exclusion of anything Euclidean, or about linear transformations.

This limitation, where basic material is separated from further reaches of the subject, has an appeal of its own. New wrinkles are discovered in heavily trodden ground, and though any points that can be made in the sparse area may perhaps be small, impacts should multiply from traffic around it.

The limitation opens opportunities, beside the accommodations and emphasis that prompted the work in the first place. There is occasional dealing with history, and in avoidance of excessive austerity—and pursuit of favourite topics—some license with the declared limitation. The contents of the book is outlined further in the introduction, where it is exposed as an essay motivated by enlargements around a number of points. Among items that may be in some way new, scattered in the work though not always with identification as such, are the following.

First is the contribution from Albert Tucker of the 'rank reduction' step, which reduces rank by one. Like the 'elementary operations' of the textbooks, it is simple but, as I discovered, has a complete scope for basic linear algebra, both for proving theorems, and computations, solving equations and so forth. Nonetheless in his judgement, and mine, it was new. It has a resemblance to his LP pivot operation, from the symmetry as between rows and columns, and the applicability to linear algebra which he had before elaborated, though it is altogether simpler. As marvellous as the idea itself is how it could have been missed for so long.

To the 'elementary operations' method of the textbooks for doing linear algebra, not dealt with here, Tucker added a method with his 'pivot operation'. I have pursued another method based on the 'linear dependence table', and now another based on Tucker's 'rank reduction'. The last two approaches are my favourites.

One hardly ever calculates the determinant from its algebraical formula, or ever uses Cramer's rule to solve equations. Nonetheless these are classic features in the theory of linear dependence, not to be abandoned. The *determinant* is introduced here in a completely unusual upside-down fashion where *Cramer's rule* comes first. Given $n + 1$ vectors of order n of which n are independent, one is a unique combination of the others, where the coefficients are rational functions of elements. By consideration of exchange between vectors, these coefficients must have the familiar quotient form provided by Cramer's rule, in terms of a function that must have the Weierstrass characteristic properties, thus identifying it with the determinant.

I also deal with what I believe to be a completely new idea, of the *alternant* (not the 'alternant' that has an occurrence in Muir's *History of Determinants*), a function associated with the affine space the way the determinant is with the linear space, with $n + 1$ vector arguments, as the determinant has n. Just as the determinant's non-vanishing is the condition for linear independence, and for having a base in the linear space, non-vanishing of the alternant is the condition for affine independence, or for having a regular simplex, making a base for *affine* (or barycentric) coordinates. Then for these we find a 'rule' which is an unprecedented exact counterpart of Cramer's rule for linear coordinates, which we call the *affine Cramer rule*, where the alternant takes on the role of the determinant.

These are among the more distinct or spectacular items for possible novelty, or unfamiliarity. Others, with or without some remark, may be found scattered in different places.

An account of permutations required for determinants is in an appendix. Here we also give a treatment of the order of last differences used to locate submatrices, or elements of derived matrices, the *systèmes dérivés* of the Binet-Cauchy Theorem. This is offered for its own interest. With the kth derived matrix, of an $m \times n$ matrix, we are in effect dealing with combinations k at a time of m or n elements, the rows or columns, ordered by last differing elements. Though the formation of the elements of a derived matrix as determinants of submatrices is straightforward, it gives rise to the question of how one proceeds backwards, from an element of the derived matrix to the submatrix of the original with which it is associated.

Giving an algorithm realization in the form of a computer program is a quite fitting conclusion to dealing with it. A collection of listings of such programs, together with demonstrations by illustrative examples, is at the end of the book. The programs are rather for browsing, especially by readers who might like to write their own. I use MS QuickBASIC which has both an interpreter, which used to come free with DOS so just about everyone has it, and a compiler. It is the most generally available and readable of languages, and has other qualities making it, in my opinion, most suitable for the use it has in the book. While the linear algebra programs are based straightforwardly on algorithms developed in the text, certain of those to do with listing permutations and combinations are without any such readiness and perhaps have a further interest.

Contents

Chapter 5 Replacement

Chapter 6 Linear Equations

Chapter 7 Determinants

Chapter 8 Determinants and Matrices

Chapter 9 Quadratic Forms

Appendix

BASIC Programs

Bibliography

BASIC Programs

6 Combinations

Acknowledgments

My thanks go to Elena Carsi for assistance in preparations, as also to Simon Smith, author of the wordprocessor $\mathbb{EXP}$ used for making this book.

I also thank Ali Doğramaçi, of Bilkent University, Ankara, for hospitality at his university during a period of the work, before its conclusion at the University of Siena.

Linear Dependence

Theory and Computation

Introduction

The subject which often used to be called something like Determinants and Matrices received a new dress when it became Linear Algebra, settling it as a branch of the more abstract subject, finite dimensional and with the real numbers, or perhaps complex numbers, as scalars. Paul R. Halmos with his *Finite Dimensional Vector Spaces* (1948) brought that shift into the teaching; so did H. L. Hamburger and M. E. Grimshaw, in *Linear Transformations in n-dimensional Vector Space* (1951), though they were looking forward to getting into Hilbert space. It is now often called classical linear algebra, to distinguish it from combinatorial linear algebra, which deals with linear inequalities and convexity and has linear programming (LP) as a topic.

This work deals mainly with the start of classical linear algebra. We follow a scheme based on the primitive idea of a *linear dependence table*, which provides the coefficients by which certain elements are combinations of certain others, the generators being associated with rows, and combinations of these with columns. Any matrix can immediately be regarded as a linear dependence table, showing its columns (or by transposition also its rows) as combinations of the fundamental base vectors. Then there is the *replacement operation*, by which a generated element takes the place of a generator. The operation is identified by the pivot element, in the row of the generator to be replaced and the column of the combination that will replace it. Any non-zero element in the table can be a pivot, representing a possible replacement. The key *Replacement Theorem* is that *the span and independence of generators is unaffected by replacement*. The Steinitz Exchange Theorem, and then the fundamental Dimension Theorem, are corollaries. A contemplation of the scheme which results following a *maximal replacement*, possibly with the pivots restricted to a subset of columns, produces at once both theoretical results and computational algorithms.

In David Gale on *The Theory of Linear Economic Models* (1960), I came across two exercises (1−2, p. 129) which suggested this scheme. Usually we have the reduction of a matrix by elementary operations. The replacement operation is in fact carried out by elementary operations, on the rows of the table. But still, here definitely is another point of view. Since any vector is a combination of fundamental base vectors with coefficients given by its own elements, there is the accident that any matrix can be regarded as a linear dependence table showing its columns as linear combinations of generators associated with the rows. Then we have the entire theory of simultaneous linear equations, and of matrix rank, and inversion, joined with computational algorithms, by application of the Replacement Theorem to a maximal replacement.

One can see here a lesson for linear algebra, that comes also from linear programming and George B. Dantzig. He showed how the Simplex Algorithm was not just for computation but, by consideration of the ways it can terminate, also a means for proof of the main theoretical results about linear programming. Albert W. Tucker accomplished this again, in a manner that at the same time gave the essential symmetry, or duality, in linear programming its full exposure.

Our interests are touched by Albert Tucker also in his treatment of linear equations, which again brings theory and computation together. He pursued this following the practice he developed for linear programming, with *pivot operation* on an LP *tableau,* which provides an implementation of the Simplex Algorithm of George Dantzig. Beside arithmetical economy, its peculiar ambiguity between primal and dual goes to the centre of that subject and its duality principle

The scheme with a *pivot operation* on a *tableau,* and a peculiar ambiguity representing a symmetry, or duality, as between rows and columns, is associated especially with Tucker. It fits linear programming, with its duality concept, in an essential way. I also heard Professor Tucker lecture on the application to basic linear algebra, parallel to that just described, bringing together theoretical results with computational algorithms. We shall see how these two schemes are connected.

When a generator is replaced, it is lost and the operation cannot be reversed, unless the generator was included among the elements generated. Hence we consider the *extended table* where the generators are so included. But then we know the columns associated with generators intersect the rows in a unit matrix, so without loss we can form the *condensed table* where these are omitted. This is the *tableau* of Tucker, and the pivot operation produces one tableau from another *reversibly,* with a swap between labels of rows and columns, instead of replacing a row label by a column label. We give an exposition of this connection, and of the duality, or peculiar ambiguity in regard to rows and columns, of the pivot operation. With Tucker, instead of explicit connection with a linear dependence table, or even just having symbolic labels for recording swaps, the rows and columns are associated with variables, base and non-base, or *vice versa* if you are thinking of the dual, and having a swap, ambiguously for primal or dual, when there is a pivot operation. We also give an exposition of this scheme, on which the LP application is based.

Any multiple replacement, or pivot, is identified with a single block replacement where, instead of a pivot element, there is a *pivot block* given by a regular square submatrix in the original table, or the tableau, replaced in the end by its inverse. This shows the principle for the efficient application of pivot operations to matrix inversion pursued by Tucker.

In a meeting with Albert Tucker in 1987, by a brief remark he conveyed the idea for a simple, and unfamiliar, algorithm. In present terms, this is the procedure for *rank reduction* of a matrix. It has perfect symmetry between rows and columns, as to be expected from Tucker. It seemed a distinctly new wrinkle in the heavily trodden ground.

Of course after Gauss[1] nothing belonging to this area could be quite new. All the same, there seemed some novelty, and there is the following from George Polya in *How to Solve it*: "Try to treat symmetrically what is symmetrical ... ".

The rank reduction step is close to the formula for the *critical decomposition* of a matrix, which I knew and used, and had not come across elsewhere. I found he had this also, with the belief that it had still to appear in print, which impression I also had.

With any matrix *a,* by taking any non-zero element and subtracting the product of its column and row, divided by it, the rank is reduced by 1. This is the *rank reduction* step. By repeating it *r* times to produce the null matrix, and recording the pivot rows and columns, the rank *r* is discovered, and so also are row and column bases *u* and *v,* intersecting in a critical submatrix *p.* From symmetry of this process between rows and columns comes yet another proof that row rank equals column rank. The product of the

---

[1] Or after 3rd century BC to allow for early Chinese knowledge. See Carl B. Boyer, *A History of Mathematics*, 1968, pp. 218-9.

pivot elements provides the determinant of p. The decomposition formula just cited is that

$$a = vp^{-1}u.^2$$

For a generalization, of both this and the rank reduction step, if p is regular $s \times s$, then

$$b = a - vp^{-1}u.$$

is of rank $r - s$.

For the case of a regular square matrix, we have $p = a$, so here is a way of evaluating the determinant. It should be connected with the *pivotal condensation* method accounted by A. C. Aitken (1942, pp. 45-7), "ascribed to Chiò (1853), but virtually used by Gauss more than forty years earlier in evaluating symmetric determinants". Now the method is expressed with proper simplicity, and symmetry, inherited from the rank reduction procedure.

Tucker had brought attention earlier to what he called the Gauss-Grassmann decomposition of a matrix, of rank r, into a sum of r matrices of rank 1, the *primitive matrices* of Grassmann (1944), or *dyads* of Gibbs (1884). He proposed an algorithmic scheme, and contemplating it found "This key piece of elementary matrix algebra (1) unites computation and theory, (2) uses classical tools, and (3) shows fundamental duality between rows and columns."[3] A later modification brought out the symmetry further.[4]

The decomposition formula enables the most immediate special features for symmetric matrices to be shown, as in §9.1, with extra simplicity. Then there are results applicable to quadratic forms, and to the bordered matrices dealt with when these are linearly restricted. For instance the test for a positive definite matrix, usually stated in terms of determinants, has a more practical form that requires the pivots in a rank reduction procedure, taken from the diagonal, all to be positive, and there is a similar method for when there are linear restrictions. The methods settle with brevity questions about quadratics, and this has led to a fairly extensive account, including elaborations around Finsler's theorem involving matrix pencils.

From dealing much with partitioned matrices, it seemed suitable to include 'double determinants', found in §7.5, though applications

---

[2] With whatever critical p, the matrix $v'au'$ is regular, and so one may consider $a^+ = u'(v'au')^{-1}v'$ which, immediately, is independent of the particular p and so a single-valued function of a. This is a formula for the Moore-Penrose inverse, discovered also by L. B. Willner (1967).

[3] Tucker (1978).

[4] Tucker (1980).

may be far from this book.[5]

As well known and apparent again from this book, one hardly ever calculates the determinant from its algebraical formula, or ever uses Cramer's rule to solve equations. Nonetheless these are classic features in the theory of linear dependence, not to be abandoned. The *determinant* is introduced here in an upside-down fashion where *Cramer's rule* comes first. Given $n + 1$ vectors of order n of which n are independent, one is a unique combination of the others, where the coefficients are rational functions of elements. By consideration of exchange between vectors, these coefficients must have the familiar quotient form provided by Cramer's rule, in terms of a function that must have the Weierstrass characteristic properties, so identifying it with the determinant. This completely unusual procedure has its advantage.

We also bring in the idea of the *alternant*[6], a function associated with the affine space the way the determinant is with the linear space, with $n + 1$ vector arguments, as the determinant has n. Just as the determinant's non-vanishing is the condition for linear independence, and for having a base in the linear space, non-vanishing of the alternant is the condition for affine independence, or for having a regular simplex, making a base for *affine* (or barycentric) coordinates. Then for these we find a 'rule' which is an exact counterpart of Cramer's rule for linear coordinates, which we might call the *Affine Cramer Rule*, where the alternant takes on the role of the determinant.

The terms from abstract algebra that bear on the linear space are reviewed in the first chapter. With the usual economy which we had perhaps first from Schreier and Sperner (1920-30s) and then Halmos (1948), the space is dealt with abstractly until we have the finite dimension; and even then, unless faced with an actual matrix, we may put matters in terms of points in the space, though they could be coordinate vectors, or simply vectors.

An account of permutations required for determinants is in an appendix. Here we also give a treatment of the order of last differences used to locate submatrices, or elements of derived matrices, the *systèmes dérivés* of the Binet-Cauchy Theorem. This is offered for its own interest or value rather than another importance. With the kth derived matrix, of an $m \times n$ matrix, we are

---

[5] Egerváry (1960) gives one to do with electrical circuits; another is where it provides a route to the rational canonical form of a matrix, as shown by Afriat (1973).

[6] This is not the 'alternant' that has an occurance in some volume of Muir's *History of Determinants*.

in effect dealing with combinations k at a time of m or n elements, the rows or columns, ordered by last differing elements. Though the formation of the elements of a derived matrix as determinants of submatrices is straightforward, it gives rise to the question of how one proceeds backwards, from an element of the derived matrix, to the submatrix of the original with which it is associated.

Giving an algorithm realization in the form of a computer program is an instructive conclusion to dealing with it. A collection of listings of such programs, together with demonstrations by illustrative examples, is at the end of the book.

1
Matrices

1.1 Matrices and vectors

The matrices of order $m \times n$ over a field K are denoted K_n^m. Any matrix $a \in K_n^m$ has elements

$$a_{ij} \in K \ (i = 1, \dots , m; \, j = 1, \dots , n),$$

ordered in a rectangular array with m rows and n columns, m being the *row order* and n the *column order* of an $m \times n$ matrix. The array of elements defines the matrix, so

$$a = \begin{bmatrix} a_{11} & \dots & a_{1n} \\ \dots & \dots & \dots \\ a_{m1} & \dots & a_{mn} \end{bmatrix}.$$

The $m \times 1$ and $1 \times n$ matrices in particular are denoted

$$K^m = K_1^m, \quad K_n = K_n^1.$$

Any $x \in K^m$, $y \in K_n$ are represented by arrays

$$x = \begin{bmatrix} x_1 \\ \vdots \\ x_m \end{bmatrix}, \, y = \begin{bmatrix} y_1 \dots y_n \end{bmatrix}$$

with a single column or row; x is distinguished as a *column vector*, of *order m*, and y as a *row vector*.

The elements of $a \in K_n^m$ have rows

$$a_{(i} \in K_n \ (i = 1, \dots , m)$$

and columns

$$a_{j)} \in K^m \ (j = 1, \dots , n),$$

where

$$a_{(i} = \begin{bmatrix} a_{i1} \dots a_{in} \end{bmatrix}, \quad a_{j)} = \begin{bmatrix} a_{1j} \\ \vdots \\ a_{mj} \end{bmatrix}.$$

A matrix can also be displayed through its rows, or its columns:

$$a = \begin{bmatrix} a_{(1} \\ \vdots \\ a_{(m} \end{bmatrix}$$

is the *row partition* form of a, and

$$a = [\, a_{1)} \ \ldots \ a_{n)} \,]$$

the *column partition* form. In these ways a matrix is represented by a set of vectors, provided either by its rows or by its columns.

The *transpose* of a matrix a is obtained by interchanging rows and columns; denoted a', it is the matrix with elements

$$a'_{ij} = a_{ji}.$$

If a is $m \times n$, then a' is $n \times m$, and

$$(a')_{(k} = (a_{k)})',$$

or row k of the transpose a' is obtained by transposing column k of a.

A *square matrix* has the same row and column order, so the matrices K_n^n, with the same number n of rows and columns, are the square matrices of order n. The *diagonal* of such a matrix a is described by its elements a_{ij} for which $i = j$, so there are n *diagonal elements* a_{ii} $(i = 1, \ldots , n)$. A *diagonal matrix* a has all its non-diagonal elements zero, so $a_{ij} = 0$ for $i \neq j$.

A *symmetric matrix* is identical with its transpose, so it is necessarily a square matrix. Hence for any symmetric matrix, $a = a'$, that is, $a_{ij} = a_{ji}$.

1.2 Submatrices

For $a \in K_n^m$, and subsets

$$i = (i_1, \ldots , i_r), \; j = (j_1, \ldots , j_s),$$

of r rows and s columns, where

$$1 \leq i_1 < \cdots < i_r \leq m, \; 1 \leq j_1 < \cdots < j_s \leq n,$$

we have an $r \times s$ matrix

$$a_{ij} = \begin{bmatrix} a_{i_1 j_1} & \cdots & a_{i_1 j_s} \\ \cdots & \cdots & \cdots \\ a_{i_r j_1} & \cdots & a_{i_r j_s} \end{bmatrix},$$

with elements drawn from the elements of a. This is the *submatrix* of a from those rows and columns. We also use the notation $a_{(i}$ for the $r \times n$ submatrix of a on rows in the set i, and

$$a_{[i} = a_{(i'}$$

for the $(m - r) \times n$ submatrix of rows in the complementary subset i'; and similarly with columns.

In the case of leading sets

$$i = (1, \ldots, r), \quad j = (1, \ldots, s),$$

so with complements

$$i' = (r+1, \ldots, m), \quad j' = (s+1, \ldots, n),$$

the matrix can be displayed in the partitioned form

$$a = \begin{bmatrix} a_{ij} & a_{ij'} \\ a_{i'j} & a_{i'j'} \end{bmatrix},$$

as a 2×2 array of submatrices. The a_{ij}, $a_{i'j'}$ are *complementary submatrices*, being drawn from complementary sets of rows and columns. Similarly, given any partitions of the m rows and n columns into consecutive subsets of $m_1, \ldots, m_h$ and $n_1, \ldots, n_k$ elements, where

$$m_1 + \cdots + m_h = m, \quad n_1 + \cdots + n_k = n,$$

we have the matrix in the *consecutively partitioned* form

$$a = \begin{bmatrix} A_{11} \ldots A_{1k} \\ \cdots \cdots \cdots \\ A_{h1} \ldots A_{hk} \end{bmatrix}$$

as an $h \times k$ array of submatrices of a, where A_{rs} is of order $m_r \times n_s$.

A square matrix has a *symmetric partition* when the same partition is applied to the rows and columns, so the diagonal blocks are all square. In the case

$$a = \begin{bmatrix} A_{11} & A_{12} \\ A_{21} & A_{22} \end{bmatrix}$$

the blocks A_{11}, A_{22} are square submatrices of a of order n_1, n_2; and A_{12}, A_{21} are of order $n_1 \times n_2$, $n_2 \times n_1$. A *principal submatrix* of a square matrix is a submatrix, necessarily square, drawn from the same set of rows and columns. The diagonal blocks A_{11}, A_{22} in the symmetric partition of a are principal submatrices; and, as such, they are complementary.

1.3 Matrix operations

There are three basic operations with matrices.

(I) *scalar multiplication*, the multiplication of a matrix by a scalar: any $a \in K_n^m$ and $t \in K$ determine $at \in K_n^m$, where

$$(at)_{ij} = a_{ij} t,$$

in other words, every element of a is multiplied by t.

(II) *matrix addition*: $a, b \in K_n^m$ determine $a + b \in K_n^m$, where

$$(a + b)_{ij} = a_{ij} + b_{ij}.$$

To form the sum of two matrices, they must be of the same order, making them *conformable for addition*, and then corresponding elements are added.

(III) *matrix multiplication*: $a \in K_q^p$, $b \in K_r^q$ determine $ab \in K_r^p$, where

$$(ab)_{ij} = \sum_k a_{ik} b_{kj}.$$

Hence to form the product ab, the column order of a must equal the row order of b, making a, b *conformable for multiplication*, in that order. Then

$$(ab)_{ij} = a_{(i} b_{j)} \qquad\qquad (i)$$

so the ij-th element of ab is obtained by multiplying row i of a with column j of b. The row order of ab is that of a, and the column order that of b. Also we have

$$(ab)_{(i} = a_{(i} b, \quad (ab)_{j)} = ab_{j)}, \qquad\qquad (ii)$$

so row i of ab is obtained by multiplying just row i of a with the matrix b, and correspondingly for a column.

With reference to the notation for submatrices in §2, these same formulae (i) and (ii) are applicable when individual rows and columns i, j are replaced by subsets $\boldsymbol{i}, \boldsymbol{j}$.

Evidently,

$$(ab)' = b'a',$$

or the transpose of a product is the product of transposes of the factors taken in reverse order; hence, for several factors,

$$(ab \cdots c)' = c' \cdots b'a'.$$

In the way of syntax, we usually represent scalar multiplication of a column vector as a multiplication from the right, and that of a row vector as from the left, as if the scalar were a 1×1 matrix.

For $a \in K_n^m$, $x \in K_n$ and $b \in K^m$, the matrix relation

$$ax = b,$$

in terms of elements, is that

$$\sum_j a_{ij} x_j = b_i \quad (i = 1, \dots, b).$$

This states the *simultaneous equations*

$$a_{(i} x = b_i \quad (i = 1, \dots, m),$$

and, alternatively,

$$a_{1)} x_1 + \cdots + a_{n)} x_n = b,$$

that *the vector b is a linear combination of the columns of the matrix a, with coefficients given by the elements of the vector x.*

Consider a, b of order $p \times n$, $n \times q$ and so conformable for multiplication with product $c = ab$, of order $p \times q$. We say a, b are *conformably partitioned* if the same partition applies to the columns of a as to the rows of b. If A, B are such partitioned representations of a, b the blocks A_{rt}, B_{ts} have some orders $p_r \times n_t$, $n_t \times q_s$ and so are conformable for multiplication, with products of order $p_r \times q_s$, so conformable for addition for all t. Hence we may form the matrices

$$C_{rs} = \sum_t A_{rt} B_{ts},$$

these being `elements' of the matrix product AB, formed as if the arrays A, B were matrices whose elements are matrices. The *theorem on the multiplication of partitioned matrices* asserts that C is a partitioned representation of c. In this, the partitions in the rows and columns of c are those in the rows of a and the columns of b, respectively. With the obvious similar rules for scalar multiplication and matrix addition, the theorem offers that partitioned matrices, provided they are conformable, can be treated for matrix operations as if they were matrices whose elements are matrices.

Evidently, matrix addition is commutative,

$$a + b = b + a,$$

but not matrix multiplication. Both operations are associative,

$$a + (b + c) = (a + b) + c, \quad a(bc) = (ab)c.$$

Also, matrix multiplication is distributive over addition from both sides,

$$a(b + c) = ab + ac, \quad (a + b)c = ac + bc.$$

A *null matrix* has all its elements 0, and is denoted o; a *unit matrix* is a square matrix with all diagonal elements 1 and other elements 0, and is denoted 1. We have

$$a + o = a, \quad o + a = a,$$

$$a1 = a, \quad 1a = a,$$

$$ao = o, \quad oa = o.$$

The columns of $1 \in K_n^n$ are the *fundamental vectors* in K^n, and similarly with the rows as vectors in K_n. These vectors all have a single element 1 and the other elements 0.

The *powers* a^r $(r = 0, 1, \dots)$ of a square matrix a can be defined by

$$a^0 = 1, \quad a^r = aa^{r-1} \; (r = 1, 2, \dots).$$

A different definition of powers useful for certain applications is based on the modified arithmetic where PROD means SUM, and SUM means MIN. It serves for a reformulation of the shortest path algorithm of Ford and Fulkerson (1962). An account together with a BASIC program that exploits the idea, and the application to economic index numbers, is in my 1987 book.[7]

Because matrix multiplication is associative,

$$a^r a^s = a^{r+s}.$$

From this, though matrix multiplication is generally not commutative, all powers of a matrix commute,

$$a^r a^s = a^s a^r.$$

A consequence of the commutation of powers is that any polynomial identity remains an identity when the scalar indeterminate z is replaced by a square matrix a. For instance, from

$$(1 - z)(1 + z + \cdots + z^{n-1}) = 1 - z^n$$

we have

$$(1 - a)(1 + a + \cdots + a^{n-1}) = 1 - a^n,$$

which has frequent use. For example,

$$\sum_{n=0}^{\infty} a^n \rightarrow (1 - a)^{-1}$$

if and only if

$$a^n \rightarrow o \; (n \rightarrow \infty).$$

---

[7] In Chapter V.4; also the 1981 paper, which develops further the report of 1960 reproduced in the 1963 article.

1.4 Algebra

Basic terms of algebra will now have a review. Though this book is not directed towards abstract algebra, it is suitable to recognize its place there, and the common features; and to take advantage in adoption of forms and economies, as in parts of matrix theory where coordinates are put out of the way when not essential.

A *binary operation* in a set K is a mapping

$$\circ : K \times K \ \rightarrow \ K,$$

where any $a, b \in K$ determine an element $a \circ b \in K$. The operation is *commutative* if

$$a \circ b = b \circ a,$$

and *associative* if

$$a \circ (b \circ c) = (a \circ b) \circ c.$$

A set with an associative binary operation defines a *semigroup*, for example, the positive integers where $a \circ b$ is $a + b$, or ab, or $a + b + ab$, or the subsets of a given set where $a \circ b$ is $a \cup b$, or $a \cap b$. Because of the associativity, the expression

$$a_1 \circ \cdots \circ a_n$$

is well defined for a semigroup, and so are *powers*

$$a^n = a \circ \cdots \circ a.$$

An element e is a *left identity* if $e \circ a = a$ for all a; and similarly for a *right identity*. For example, the positive integers with multiplication have 1 as a left and right identity, but with addition have no identity; any semigroup with $a \circ b = b$ has every element as a left identity, but no right identity.

If left and right identities exist, they must be identical, and so also unique. For we have $e \circ f = f$ if e is an identity on the left, and $e \circ f = e$ if f is on the right, so $e = f$. There can be no other left or right identities e' or f'. For, given any, we would have $e' = f$, $f' = e$ so these would be no different from the former, or each other. By an *identity* is meant an element which is both a left identity and a right identity. If there is one, it is necessarily unique.

For a semigroup with identity e, an element a has b as a *left inverse* if $b \circ a = e$, and as a *right inverse* if $a \circ b = e$. If an element has both a left and a right inverse, these must be identical, and unique. Accordingly, if b and c are left and right inverses of a, we have

$$b \circ a = e, \quad a \circ c = e,$$

so that

$$b \circ a \circ c = e \circ c = c, \quad b \circ a \circ c = b \circ e = b$$

and hence $b = c$. Also there can be no other left or right inverses. Hence there is a unique element, the *inverse* of a denoted a^{-1}, for which

$$a^{-1} \circ a = e = a \circ a^{-1}.$$

An element is *regular* if it has an inverse, and otherwise *singular*.

A *group* is a semigroup with an identity, and every element regular.

Consider a set K now with two binary operations, called *addition* and *multiplication*, by which any $a, b \in K$ determine

$$a + b, \ ab \in K.$$

The set with these operations is a *ring* if we have a commutative group with addition, a semigroup with multiplication, and multiplication is *distributive over addition, from the left* and *right*,

$$a(b + c) = ab + ac, \quad (a + b)c = ac + bc.$$

For example the real square matrices K_n^n, with matrix addition and multiplication, form a ring.

A *field* is a ring where also multiplication is commutative, and the elements different from the identity o in the additive group form a group under multiplication, for example, the real or complex numbers. A *linear space over a field* is defined in Chapter 2. An *algebra* is a ring which is also a linear space over a field.

A *binary relation* in a set S is a subset $R \subset S \times S$. The statements xRy, $x \in Ry$ and $y \in xR$ are alternatives to $(x, y) \in R$, to assert x has the relation R to y. We may use $xRyRzR\ldots$ to mean $xRy, yRz, \ldots$ and $x, y, \ldots Rz$ to mean $xRz, yRz, \ldots$.

An *order* of a set S is a binary relation R in S with the properties of *reflexivity*,

$$x \in S \implies xRx,$$

and *transitivity*,

$$xRyRz \implies xRz.$$

An *ordered set* is a set for which an order is defined. For example, numbers ordered by magnitude, subsets of a set by inclusion, subspaces of a linear space by the subspace relation, or polynomials over a field by one being a divisor of another. A *least upper bound* in an ordered set is a binary operation $\vee$ where, if $<$ denotes the order,

$$x, y < x \vee y; \quad x, y < z \Rightarrow x \vee y < z.$$

A *greatest lower bound* $\wedge$ is defined similarly. A *lattice* is an ordered set for which a least upper bound and greatest lower bound are defined. With the above examples, we have lattices with the operations max and min, union and intersection, least common multiple and greatest common divisor.

1.5 Vector space

Linear algebra has roots in geometry or mechanics, where there is an idea of the displacement from one point in a space to another by application of a vector, the portability of vectors, and their composition by the parallelogram law of addition.

We have a *space S*, the elements of which are *points*, and a set V of operations in S, the *translation vectors*. For any $x \in V$ and $P \in S$, there corresponds a point, the *translation* of P by x, denoted $Q = P + x$. For all P, Q there exists just one x such that $P + x = Q$, the *vector* from P to Q, denoted $x = \langle P, Q \rangle$, so now

$$x = \langle P, Q \rangle \Leftrightarrow Q = P + x.$$

An assumption is that $(P + x) + y = (P + y) + x$, so the applications of vectors successively to any point are commutative. We now have a *congruence space S* with V as its *vector space*. Ordered pairs of points are *congruent* if they have the same vector,

$$(P, Q) \equiv (P', Q') \Leftrightarrow \langle P, Q \rangle = \langle P', Q' \rangle.$$

The congruence of one pair of opposite sides of a quadrilateral implies that of the other pair,

$$\langle P, A \rangle = \langle B, Q \rangle \Leftrightarrow \langle P, B \rangle = \langle A, Q \rangle.$$

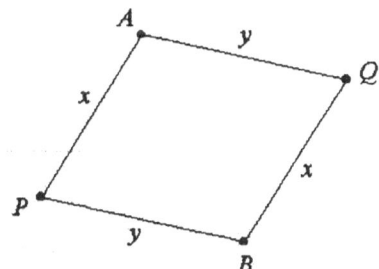

To see this, let

$$\langle P,B \rangle = y, \ \langle A,Q \rangle = y, \ \langle P,A \rangle = x.$$

Then it has to be shown that $\langle B,Q \rangle = x$, that is,

$$\langle (P+y),(P+x)+y \rangle = x.$$

But that is $(P+x)+y = (P+y)+x$, which has been assumed. For another statement, ordered pairs of points are congruent if an only if they correspond under a translation, that is,

$$(P,A) \equiv (B,Q) \ \Leftrightarrow \ B = P+t, \ Q = A+t \ \text{for some } t.$$

Hence we have

$$\langle (P+x),(Q+x) \rangle = \langle P,Q \rangle,$$

for all P,Q and x.

It is to be seen that for all $x,y \in V$ there exists a $z \in V$ such that

$$(P+x)+y = P+z,$$

for all $P \in S$. We want $z = \langle P,(P+x)+y \rangle$ to be independent of P, or what is the same, since any point of S is of the form $P+t$ for some t,

$$\langle (P+t),((P+t)+x)+y \rangle$$

is to be independent of t. But, by two applications of commutativity, this expression is

$$\langle (P+t),((P+x)+y)+t \rangle,$$

and, since correspondence by translation implies congruence, this is

$$\langle P,(P+x)+y \rangle,$$

which is independent of t.

Now z can define the *sum* $x+y \in V$ of any x, $y \in V$ with the property

$$P+(x+y) = (P+x)+y.$$

This operation is associative,

$$x+(y+z) = (x+y)+z,$$

since

$$P+((x+y)+z)$$
$$= (P+(x+y))+z = ((P+x)+y)+z = (P+x)+(y+z)$$
$$= P+(x+(y+z)).$$

Also it is commutative, $x+y = y+x$, since

$$P+(x+y) = (P'+x)+y = (P+y)+x = P+(y+x).$$

Given any $x, y \in V$ their sum can be formed as follows, according to the parallelogram law of vector addition.

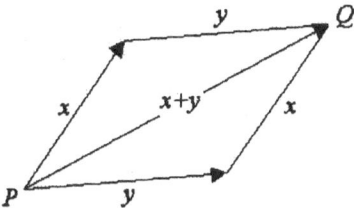

Take any point P and form the quadrilateral

$$P, \ P + x, \ P + y, \ Q$$

where

$$(P + x) + y = Q = (P + y) + x.$$

It defines a *parallelogram*, in that its opposite pairs of sides are congruent. Now $x + y = \langle P, Q \rangle$, that is, $x + y$ is the vector along the diagonal from P in the parallelogram in which P is a vertex and x, y are the vectors along sides adjacent at P. From

$$\langle P, Q \rangle = \langle P, Q \rangle$$

follows

$$\langle P, P \rangle = \langle Q, Q \rangle,$$

so the vector $o = \langle P, P \rangle$ is independent of P. It has the property $P + o = P$ for all P. Now

$$P + (x + o) = (P + x) + o = P + x,$$

so it also has the property $x + o = x$, for all x. It is the unique vector in V with this property. For, were o' another, we would have $o = o + o' = o'$. From

$$\langle P + x, Q + x \rangle = \langle P, Q \rangle$$

follows

$$P = (P + x) + t, \ Q = (Q + x) + t,$$

where

$$t = \langle P + x, P \rangle$$

is independent of P. Hence for all x there exists a t with the property that, for all P,

$$(P + x) + t = P,$$

that is,

$$P + (x + t) = P,$$

showing that $x + t = o$. If $x + y = o$ for any y, then

$$P = P + (x + y) = (P + x) + y,$$

so that $y = \langle P + x, P \rangle = t$. Hence for all x there exists a unique vector, the *negative* of x denoted $-x$, with the property

$$x + (-x) = o.$$

For any x, y the difference

$$y - x = y + (-x)$$

is such that

$$(y - x) + x = y.$$

Hence any equation $a + x = b$ has a unique solution $x = b - a$. With the operation of addition, V is therefore a commutative group.

   With an arbitrary point $O \in S$, every point $P \in S$ has an expression in the form $P = O + x$, for a unique $x \in V$, the *coordinate vector* of P with *origin* O given by $x = \langle O, P \rangle$. Relative to any origin, a one-one correspondence is now determined between the points of S and the vectors of V; and $O \in S$ corresponds to $o \in V$.

   The set V of translation vectors in S appears itself as a congruence space, with itself as its set of translation vectors, where $x + t$ is the translation of x by t, and $\langle x, y \rangle = y - x$ is the vector from x to y. It is an isomorphic image of S, the isomorphism being determined by the correspondence of $o \in V$ with an arbitrary $O \in S$. Any congruence space can therefore be considered simply through the commutative group of its translation vectors. And any commutative group appears as a congruence space, with itself as the vector space; in this way, any commutative group appears as a vector space.

   With $\mathcal{V} = V^V$, an operator $f \in \mathcal{V}$, that is, a mapping

$$f : V \rightarrow V,$$

determines an image $fx \in V$ for any $x \in V$. With V as any set, the product $fg \in \mathcal{V}$ of any $f, g \in \mathcal{V}$ is defined by $(fg)x = f(gx)$. This multiplication operation in $\mathcal{V}$ is associative, $f(gh) = (fg)h$, since

$$(f(gh))x = f((gh)x) = f(g(hx)) = (fg)(hx) = ((fg)h)x.$$

   Now further, with V as a vector space, the sum $f + g \in \mathcal{V}$ can be defined, through the addition in V, by

$$(f + g)x = fx + gx.$$

This operation of addition in $\mathcal{V}$ is commutative and associative, by the same properties for addition in V. With

$$o, 1 \in \mathcal{V}$$

such that

$$ox = o, \quad 1x = x$$

for all $x \in V$, we have

$$of = o, \quad f + o = f = o + f, \quad 1f = f = f1,$$

but not $fo = o$ ($o \in V$) unless $fo = o$ ($o \in V$). For $-f \in V$ defined by

$$(-f)x = -(fx),$$

we have

$$f + (-f) = o.$$

It has appeared that V is a *semigroup with multiplication* and a *commutative group with addition*.

An endomorphism of V is an operator $f \in V$ which preserves addition, so the image of a sum is the sum of the images,

$$f(x + y) = fx + fy.$$

Let $\mathcal{E} \subset V$ be the set of endomorphisms of V. Evidently then $\mathcal{E}$ is closed under the operations of multiplication and addition in V,

$$f, g \in \mathcal{E} \Rightarrow fg, f + g \in \mathcal{E}.$$

For $f, g \in \mathcal{E}$ implies

$$(fg)(x + y)$$
$$= f(g(x + y)) = f(gx + gy) = f(gx) + f(gy)$$
$$= (fg)x + (fg)y,$$

so $fg \in \mathcal{E}$, and

$$(f + g)(x + y) = f(x + y) + g(x + y)$$
$$= (fx + fy) + (gx + gy)$$
$$= (fx + gx) + (fy + gy) = (f + g)x + (f + g)y,$$

so $f + g \in \mathcal{E}$. Moreover for endomorphisms, multiplication is distributive over addition, from both sides,

$$f(g + h) = fg + fh, \quad (f + g)h = fh + gh.$$

For

$$(f(g + h))x$$
$$= f((g + h)x) = f(gx + hx) = f(gx) + f(hx) = (fg)x + (fh)x$$
$$= (fg + fh)x,$$

so $f(g + h) = fg + fh$, and similarly with the other side.

Now for $o, 1 \in V$,

$$o(x + y) = o = o + o = ox + oy,$$

$$1(x + y) = x + y = 1x + 1y,$$

and hence $o, 1 \in \mathcal{E}$.

With all this it has appeared that, with the addition and multiplication, $\mathcal{E}$ *is a ring.* A subring $K \subset \mathcal{E}$ which is commutative, and non-zero elements form a group for multiplication, defines a field in $\mathcal{E}$. A vector space V, together with a field K of endomorphisms, constitutes a *linear space*, with V the commutative group of *vectors*, and K the field of *scalars*. The image of $x \in V$ under $t \in K$ is denoted xt, and this defines the operation of *scalar multiplication* in V.

2
Linear Space

2.1 Linear space

A commutative group L, with group operation $+$ and identity o, is considered together with a field K, and an operation between them by which any $t \in K$, $x \in L$ determine $xt \in L$, with the properties

$$x1 = x, \quad (x+y)t = xt + yt,$$
$$x(s+t) = xs + xt, \quad x(st) = (xs)t.$$

The system formed by K and L and the operation between them with these properties defines a *linear space L over K*. Elements of L are *vectors*, and those of K are the *scalars*, and the operations are called *vector addition* and *scalar multiplication*.

The vector space $\mathfrak{R}^n$, with the usual operations defined for it, is an example of a linear space over the real numbers $\mathfrak{R}$, or a *real linear space*.

From the axioms, we also have

$$x0 = o, \quad x(-1) = -x.$$

For we have

$$x = x1 = x(1+0) = x1 + x0 = x + x0,$$

showing that, $x0 = o$, and then

$$o = x0 = x(1+(-1)) = x1 + x(-1) = x + x(-1),$$

so that $x(-1) = -x$.

2.2 Subspaces

For a linear space L over K, a *linear subspace* is formed by a subset $M \subset L$ which is a linear space with the field K of operations in L. For this, it just has to be closed under addition and scalar multiplication,

$$x, y \in M \implies x + y \in M,$$

$$x \in M, \; t \in K \; \Rightarrow \; xt \in M.$$

Then also M is a linear space over K.

Given the closure under scalar multiplication, closure under addition is equivalent to

$$x, y \in M \; \Rightarrow \; x - y \in M,$$

which is the condition for M to be a subgroup of L. Therefore the linear subspaces are the subgroups which are closed under scalar multiplication.

An equivalent condition for $M \subset L$ to be a linear subspace of L is that

$$x, y \in M, \; s, t \in K \; \Rightarrow \; xs + yt \in M,$$

and again,

$$x, y, \ldots \in M, \; s, t, \ldots \in K \; \Rightarrow \; xs + yt + \cdots \in M.$$

The expression $xs + yt + \cdots$ is a *linear combination* of elements $x, y, \ldots$ with *coefficients* $s, t, \ldots$. It is a *proper* linear combination unless all $s, t, \ldots = 0$.

Let $M < L$ state that M is a linear subspace of a linear space L. Given two linear subspaces of a linear space, one is a subspace of the other if and only if it is a subset of the other,

$$X, Y < L \; . \; \Rightarrow \; . \; X < Y \; \Leftrightarrow \; X \subset Y.$$

Hence the subspaces of a given space with the subspace relation between them form an ordered set.

The *sum* of any $X, Y \subset L$ is defined by

$$X + Y = \{ x + y : x \in X, y \in Y \}.$$

Evidently,

$$X, Y < L \; \Rightarrow \; X \cap Y, \; X + Y < L,$$

that is, any intersection or sum of subspaces is a subspace. To express that X, Y are subspaces, we denote the subspaces given by the intersection and sum by $X \wedge Y$ and $X \vee Y$, and call the sum their *join*.

The intersection of a pair of subspaces is the greatest subspace which is a subspace of both, and the join is the least subspace of which both are subspaces,

$$X \wedge Y < X, Y; \quad Z < X, Y \; \Rightarrow \; Z < X \wedge Y,$$
$$X, Y < X \vee Y; \quad X, Y < Z \; \Rightarrow \; X \vee Y < Z.$$

This makes the operations of intersection and join the greatest lower bound and least upper bound in the subspace order. Hence with these operations, and the order that is already defined, *the subspaces form a lattice.*

One subspace of L is provided by the null subspace O which has o as its only element, and another by L itself. A *proper subspace* is one different from these.

Any $X, Y < L$ such that

$$X \wedge Y = O, \quad X \vee Y = L,$$

are *complements*. It will appear that any subspace of a finite dimensional space has a complement. In that case the lattice of subspaces becomes a *complemented lattice*.

2.3 Span

For $S \subset L$, we may consider the set of all subspaces of L that contain it. This set is non-empty, since L belongs to it. Then the intersection of all these subspaces is itself a subspace, containing S and contained in every subspace that contains S, so it is the smallest subspace of L that contains S. This defines the linear closure, or the *span* of S, denoted $[S]$, so

$$[S] = \bigcap \{M : S \subset M < L\},$$

and this has the property

$$S \subset [S] < L; \ S \subset M < L \ \Rightarrow \ [S] < M.$$

For a given $M < L$, any $S \subset M$ such that $M = [S]$ is said to *span* M, or M has S as a *generator* set. A *finitely generated* linear space is one that has a finite set of generators.

Theorem 3.1 The span of any elements $x_1, \ldots, x_p$ is identical with the set of all their linear combinations,

$$[x_1, \ldots, x_p] = \left\{ \sum_i x_i t_i : t_i \in K \right\}.$$

This set of combinations obviously contains the elements, and is contained in every subspace that contains them, by the definition of subspace, so it remains to see that it is itself a subspace, every combination of combinations of the elements being again in the set of combinations. Taking

$$y_j = \sum_i x_i a_{ij}$$

as any combinations of the x's, and

$$z = \sum_j y_j \, t_j$$

as any combination of these combinations, we have

$$z = \sum_j \left(\sum_i x_i \, a_{ij} \right) t_j = \sum_i x_i \left(\sum_j a_{ij} \, t_j \right) = \sum_i x_i \, s_i,$$

which shows that this is again a combination of the x's, where

$$s_i = \sum_j a_{ij} \, t_j,$$

that is, $s = at$, where a, s and t are the matrices of coefficients. QED

With the identification just shown, we now have

$$y_1, \ldots, y_q \in [x_1, \ldots, x_p] \Rightarrow [y_1, \ldots, y_q] < [x_1, \ldots, x_p],$$

and the condition for any M to be a linear subspace can be stated

$$x_1, \ldots, x_p \in M \Rightarrow [x_1, \ldots, x_p] \subset M.$$

It can be added, the join of two subspaces $X, Y < L$ is the span of their set-union,

$$X \vee Y = [X \cup Y].$$

2.4 Direct sum

Given two linear spaces X, Y over a field K, consider elements $z = (x, y)$ in the Cartesian product $Z = X \times Y$, so $x \in X$, $y \in Y$, and operations of addition $+$ in Z, and scalar multiplication, defined by

$$(x, y) + (x', y') = (x + x', y + y')$$

and, for $t \in K$,

$$(x, y) \, t = (xt, yt).$$

Evidently Z, with these operations, is a linear space over K. Considered as such, it defines the *direct sum* of the linear spaces X, Y and is denoted $X \oplus Y$.

Theorem 4.1 $X \oplus Y$ is a linear space over K.

The mapping

$$X \oplus Y \to X,$$

where (x, y) has image x, is a homomorphism, and defines the *projection* of $X \oplus Y$ onto X. If C is any subset of $X \oplus Y$, let A, B be its projections onto X, Y. Then C is a subspace of $X \oplus Y$ if and only if A, B are subspaces of X, Y.

Theorem 4.2 The subspaces of $X \oplus Y$ are of the form $A \oplus B$ where A, B are subspaces of X, Y.

Two particular subspaces of $X \oplus Y$ are $X \oplus O$ and $O \oplus Y$. These evidently are such that their intersection is the null subspace, $O \oplus O$, and their join is $X \oplus Y$. This observation permits an entry of the following considerations.

Consider a linear space Z, and subspaces $X, Y < Z$. These are *independent* subspaces if $X \wedge Y = O$, and *complementary*, as subspaces of Z, if also $X \vee Y = Z$, so any independent subspaces are in any case complementary subspaces of their join. Since the union $Z = X \vee Y$ is described by the sum

$$X + Y = \{x + y : x \in X, y \in Y\},$$

any $z \in Z$ has an expression $z = x + y$ where $x \in X, y \in Y$. Moreover, because X, Y are independent, this expression is unique.

Theorem 4.3 If X, Y are independent subspaces, then every element $z \in X \vee Y$ has a unique expression as a sum $z = x + y$ of elements $x \in X, y \in Y$, and the mapping

$$g : X \vee Y \rightarrow X \oplus Y,$$

where $gz = (x, y)$, is an isomorphism between the join and direct sum.

For the uniqueness, were $z = x' + y'$ another such expression, we would have

$$x + y = x' + y',$$

and hence

$$x - x' = y' - y.$$

But with $x, x' \in X$ we have $x - x' \in X$; similarly $y' - y \in Y$. Accordingly, $x - x'$ is an element of X identified also as an element of Y, so it is an element of the intersection $X \wedge Y$. But this intersection is O. Hence $x - x' = o$, equivalently $x = x'$, and so also $y = y'$. QED

We now have a mapping, stated

$$g : X \vee Y \rightarrow X \oplus Y,$$

where any $z \in X \vee Y$ has image $gz \in X \oplus Y$ given by $gz = (x, y)$, for the unique $x \in X, y \in Y$ such that $z = x + y$. This mapping has an inverse,

$$g^{-1} : X \oplus Y \rightarrow X \vee Y,$$

for which $g^{-1}(x,y) = x + y$. We now have a one-one correspondence between the elements of the join $X \vee Y$ and of direct sum $X \oplus Y$. Evidently, this correspondence is moreover an isomorphism between $X \vee Y$ and $X \oplus Y$, as linear spaces over K. This makes a natural identification of $X \oplus Y$ with $X \vee Y$, for the case where X, Y are independent subspaces of some space; in this case, $X \vee Y$ can be denoted $X \oplus Y$, in order to convey that X, Y are independent.

With the Cartesian product $X \times Y$, there are the projection mappings

$$h : X \times Y \to X, \quad k : X \times Y \to Y,$$

where

$$h(x,y) = x, \quad k(x,y) = y.$$

Then mappings

$$e : X \vee Y \to X, \quad f : X \vee Y \to Y$$

are given by

$$e = hg, \quad f = kg.$$

These are endomorphisms, or linear transformations of the linear space $X \vee Y$. If 1 denotes the identity mapping, which maps every element into itself, and o the null mapping, which maps every element into o, then

$$ef = o = fe, \quad e + f = 1,$$

and also

$$e^2 = e, \quad f^2 = f.$$

These properties are not independent. Rather, all follow, for instance, just from the idempotence property $e^2 = e$, which identifies any linear transformation e as a *projector*, and $f = 1 - e$, which makes f the *complementary projector*.

The direct sum can apply to several linear spaces X_i over a field K, to produce a linear space $\bigoplus_i X_i$ which is again a linear space over K. The case where the X_i are subspaces of a given linear space, and conditions for the isomorphism between the join $\vee_i X_i$ and direct sum, will now be outlined.

Any subspaces X_i $(i = 1, 2, \dots)$ are *independent* if

$$(\vee_{i \neq j} X_i) \wedge X_j = O \quad \text{for all } j.$$

Theorem 4.4 The independence of any subspaces X_i is equivalent to the condition that, for $x_i \in X_i$,

$$\sum_i x_i = o \ \Rightarrow \ x_i = o \quad \text{for all } i.$$

It is clear that if this condition is violated, then so is the independence, and conversely.

Theorem 4.5 The independence of any subspaces X_i is equivalent to any element in their join $\bigvee_i X_i$ having a unique expression as a sum $\sum_i x_i$ of elements $x_i \in X_i$.

This is a corollary of the last theorem.

Theorem 4.6 A necessary and sufficient condition for any subspaces
$$X_i \ (i = 1, 2, \dots)$$
to be independent is that
$$(\bigvee_{i<j} X_i) \wedge X_j = O \text{ for all } j > 1.$$

The proof is by induction on the number of subspaces.

2.5 Factor space

Given a subspace $M < L$, and any $x \in L$, let
$$x + M = \{x + z : z \in M\}.$$
Then L/M can denote the set of sets of this form, so $X \in L/M$ means $X = x + M$ for some $x \in L$. Let the congruence $x \equiv y$ (mod M) be defined by the condition $x - y \in M$.

Theorem 5.1 $x \equiv y \, (\mathrm{mod}\, M) \iff x + M = y + M.$

For if $x - y = z$, where $z \in M$, then, for any $a \in M$,
$$x + a = y + z + a = y + b,$$
where $b = z + a$, so $b \in M$. This shows that
$$x + M \subset y + M.$$
Similarly there is the opposite inclusion, and hence the identity. Conversely, if we have the identity, then, since $o \in M$, we have
$$x = y + z, \text{ for some } z \in M,$$
and hence the congruence. QED

Theorem 5.2 $x \equiv y \, (\mathrm{mod}\, M), \ x' \equiv y' \, (\mathrm{mod}\, M)$

$$\Downarrow$$
$$x + x' \equiv y + y' \pmod{M}.$$

For if
$$x - y = z, \ x' - y' = z' \text{ where } z, z' \in M,$$
then
$$(x + x') - (y + y') = (x - y) + (x' - y') = z + z'.$$
But now $z + z' \in M$, whence then conclusion.

This theorem provides a way for defining addition in L/M. Let $X, Y \in L/M$, so that
$$X = x + M, \ \ Y = y + M,$$
for some $x, y \in L$. We will see that $X + Y = Z$, where
$$Z = (x + y) + M,$$
is independent of the particular x, y.

To see this, suppose also
$$X = x' + M, \ \ Y = y' + M,$$
for some other $x', y' \in L$ and let
$$Z' = (x' + y') + M.$$
We have to show that $Z = Z'$. By Theorem 5.1,
$$x \equiv x' \pmod{M}, \ \ y \equiv y' \pmod{M},$$
and then by Theorem 5.2,
$$x + y \equiv x' + y' \pmod{M},$$
whence, again by Theorem 5.1, $Z = Z'$. QED

Theorem 5.3 For $M < L$, any $X, Y \in L/M$ determine a unique sum $X + Y \in L/M$, given by
$$X + Y = (x + y) + M,$$
for any $x, y \in L$ such that
$$X = x + M, \ \ Y = y + M.$$

The following enables a scalar multiplication to be defined for L/M which, with the addition, will make it a linear space over K.

Theorem 5.4 $\ x \equiv y \pmod{M}, t \in K \ \Rightarrow \ xt \equiv yt \pmod{M}.$

Let $X \in L/M$, so that $X = x + M$ for some $x \in L$, and let $t \in K$. We want $Xt = Z$, where

$$Z = (xt) + M,$$

to be independent of the particular x. Suppose

$$X = x' + M,$$

so $x' \equiv x \pmod{M}$, and let

$$Z' = (x't) + M.$$

Then from $x \equiv x'$, by Theorem 5.4, we have $x't \equiv xt$, and hence, by Theorem 5.2, $Z' = Z$. QED

Evidently, with the now defined addition and scalar multiplication, L/M is a linear space over K. As such, for any $M < L$, this defines the *factor space L/M*.

With L considered as an additive group, and $M < L$ as a subgroup, the elements of L/M are the cosets which are elements of the factor group L/M, addition there being defined as here for the factor space.

The factor space elements can also be seen as describing a family of parallel affine subspaces, or linear manifolds.

2.6 Dual space

For a linear space L over K, a *linear form* defined on L is a mapping

$$u : L \rightarrow K,$$

of L into the scalar field, with the properties

$$u(x + y) = ux + uy,$$

for any $x, y \in L$, and for any $t \in K$,

$$u(xt) = (ux)t.$$

In other words, this is a homomorphism of L into K, with K considered as a linear space. Let L' be the set of such linear forms. Addition for any elements $u, v \in L'$ is defined by

$$(u + v)x = ux + vx,$$

and scalar multiplication of an element, with any $t \in K$, by

$$(tu)x = t(ux).$$

From its definition, and with these operations, L' *is a linear space over K;* as such, it is the *dual* of the linear space L.

Theorem 6.1 L' is a linear space over K.

Now L' being a linear space, we can consider L'', its dual. The elements of L'' are linear forms on L'. But from the definition of the operations in L', every element x of L provides such a linear form on L', and so may be mapped onto a point of L''. This *natural mapping* of L to L'' is a homomorphism. An issue is whether it is one-one, and so an isomorphism between L and its image in L''. For this it is required that

$$ux = uy \text{ for all } u \implies x = y,$$

equivalently,

$$ux = 0 \text{ for all } u \implies x = o.$$

Hence this can be affirmed under the following condition:

$$x \neq o \implies ux \neq 0 \text{ for some } u.$$

Theorem 6.2   For the natural mapping of L'' in L to be an isomorphism, it is necessary and sufficient that, for any $x \in L$, $x \neq o$, there exists $u \in L'$ such that $ux \neq 0$.

It will be obvious, in Chapter 4 where we assume the finite dimension, that this holds for a finite dimensional space. With the affirmation, L can properly be identified with the subspace of L'' which is its image in the mapping, so now $L < L''$. A *reflexive* linear space is one for which $L = L''$. It will be seen that finite dimensional spaces have this property.

2.7 Annihilators

The *annihilator* of any subset S of a linear space L is the subset S° of the dual space L' of elements u such that $ux = 0$ for all $x \in S$,

$$S^\circ = \{u \in L' : x \in S \implies ux = 0\}.$$

Theorem 7.1   The annihilator of any subset of L is a subspace of L',

$$S^\circ < L' \text{ for all } S \subset L.$$

We say $u \in L'$ *annihilates* $S \subset L$ if

$$x \in S \implies ux = 0.$$

Evidently, if u, v annihilate S, then so do $u + v$ and su, for any $s \in K$. Hence $S^\circ \subset L'$ is closed under addition and scalar multiplication, and so a subspace of L'.

In particular,

$$O^\circ = L', \quad L^\circ = O,$$

as is immediate from the definitions, in either case O denoting the null subspace, containing the null element o alone, in L or in L'. With reference to the condition in Theorem 6.2, the following can be added.

Theorem 7.2 Subject to the condition for natural isomorphism between L and L'', if $S \subset L$ contains a non-null element, then S° is a proper subspace of L'.

For a consequence, given that M is a subspace of L, we have that M is a proper subspace of L if and only if M° is a proper subspace of L'.

When we have L as a finite dimensional space, in Chapter 4, it will be possible to take up issues about annihilators involving dimension.

2.8 Direct product

The direct product $X \otimes Y$ is in some way a counterpart of the direct sum $X \oplus Y$, though its definition is without the same directness or immediacy. There would be simplification did we have the finite dimension, which is postponed to Chapter 4, and chose to rely on a use of coordinates.

With $U = X'$, $V = Y'$, let $Z = (X, Y)$, $W = (U, V)$ denote Cartesian products, so any $z \in Z$, $w \in W$ have the form $z = (x, y)$, $w = (u, v)$. These are going to be points in a linear space. But how we make Z, or W, into a space depends on how we operate in them.

For the direct sum, we should consider

$$wz = ux + vy,$$

and if we take a sum or scalar multiple of expressions with the form on the right, we obtain an expression with the same form, and so nothing new. A simplicity with the direct sum is associated with this arrangement. When Z is to be the direct sum, and we take W also as a direct sum, we obtain W as the dual of Z. To think of the direct product, we take instead

$$wz = ux \cdot vy.$$

However, sums and scalar multiples of expressions on the right, with different w (or alternatively, with different z) are not expressible in just that form. We just have a sum of expressions with that form. The direct product is defined so as to included these, and the issue now is the formal statement.

A feature about expressions on the right, and sums of these for different w, is the bilinearity in x and y. One might therefore consider proceeding with reference to the linear space of bilinear forms on Z, and make its dual the direct product space, as proposed by Halmos (1948, and 2nd edition, 1958). There is also the symbolic approach followed by Jacobson (1953). With this, the properties of the direct product operation on elements are laid out, and $wz = ux \cdot vy$ becomes stated as

$$(u \otimes v)(x \otimes y),$$

and we have expressions $x \otimes y$ as points of $X \otimes Y$, and similarly with $U \otimes V$. For the $\otimes$ operation in Z,

$$(xt) \otimes y = x \otimes (yt),$$
$$(x_1 + x_2) \otimes y = x_1 \otimes y + x_2 \otimes y,$$
$$x \otimes (y_1 + y_2) = x \otimes y_1 + x \otimes y_2,$$

and the same for W.

When we have the finite dimension, in Chapter 4, with X and Y having dimensions m and n, just as elements of the direct sum $X \oplus Y$ can be represented as vectors with $m + n$ elements x_i and y_j, so in a similar way the elements of the direct product can be represented as vectors with mn elements $x_i y_j$, and the symbols $e_i \otimes f_j$ just become labels for base elements in $X \otimes Y$.

2.9 Algebra of subspaces

The subspaces of a given linear space L are ordered by the subspace relation $X < Y$, which, as with the set inclusion relation $X \subset Y$ from which it comes, has the properties:

$$X < X, \quad \text{reflexivity}$$
$$X < Y < Z \implies X < Z, \quad \text{transitivity}$$
$$X < Y, \ Y < X \implies X = Y, \quad \text{asymmetry.}$$

Also there are the join and intersection operations, by which any $X, Y < L$ determine $X \vee Y$, $X \wedge Y < L$. These are the unique least upper and greatest lower bounds of X, Y in the subspace order, having the properties

$$X, Y < X \vee Y; \ X, Y < Z \Rightarrow X \vee Y < Z,$$

$$X, Y > X \wedge Y; \ X, Y > Z \Rightarrow X \wedge Y > Z.$$

Hence the subspaces of a given space, as an ordered set with such operations, form a lattice. The null O, the subspace with the single element o, and unit $I = L$ are universal lower and upper bounds, being such that

$$O < X < I$$

for every X of the lattice, and so

$$X \wedge O = O, \ X \vee O = X,$$

$$X \wedge I = X, \ X \vee I = I.$$

The operations of the subspace lattice have a further property which makes it a *modular lattice* (or Dedekind lattice). This is shown in the following theorem.

Theorem 9.1 $\quad X > Y \Rightarrow X \wedge (Y \vee Z) = Y \vee (X \wedge Z).$

We have

$$X < Y \Rightarrow X \wedge Z < Y \wedge Z,$$

and $Y, Z < Y \vee Z$, so

$$(X \wedge Y) \vee (X \wedge Z) < X \wedge (Y \vee Z)$$

holds unconditionally. With $X > Y$ we have $X \wedge Y = Y$, so this becomes

$$Y \vee (X \wedge Z) < X \wedge (Y \vee Z),$$

and to conclude the required identity we just have to show the opposite relation, subject to $X > Y$.

With any element $e \in X \wedge (Y \vee Z)$, we have $e = x$ where $x \in X$, and also $e = y + z$ where $y \in Y$, $z \in Z$. Therefore

$$z = x - y \in X \vee Y = X,$$

so that $z \in X \wedge Z$, and hence

$$e = y + z \in Y \vee (X \wedge Z).$$

QED

With $X > Y$, we have $Y = X \wedge Y$, and so the conclusion here can be put just as well in the distributive form

$$X \wedge (Y \vee Z) = (X \wedge Y) \vee (X \wedge Z).$$

However, this formula is not valid without the condition $X > Y$. For instance, with the example of Jacobson (1953, p. 27), let X and Y be spanned by independent x and y, and Z by $x + y$, so $Y \vee Z$ is spanned by the pair x, y and now

$$X \wedge (Y \vee Z) = X.$$

On the other hand $X \wedge Y$, $X \wedge Z = O$ so that

$$(X \wedge Y) \vee (X \wedge Z) = O.$$

2.10 Affine space

A geometrical figure can stand apart from the way it is represented, for instance with reference to an arbitrary origin. Points, lines, planes and other linear objects could be regarded without notice of such reference, but this may be spoilt when they are brought into connection with a structure like the linear space, where subspaces are all through the origin. Skew lines, without a point in common, cannot be represented as linear subspaces, but with an affine space they are affine subspaces. This is not far from the linear space, but it provides a better view for ordinary geometry. There has been a drift in the terminology, and what used to be called a vector space is now a linear space, the matter central to linear algebra. The earlier usage fitted geometry and mechanics, and would make a linear space what is now an affine space, though the affinity is specific, and linear. Now instead of a linear space and the associated vector space of displacements in it, we have an affine space and the related linear space. This may not fit so well, but is how, perhaps by pressure from algebra, the terminology seems to have settled.

There is to be a space S, and subspaces with a linearity characteristic. A first representative of linearity is the line, any one determined by two points. Then for a subspace, with any pair of points it should contain the entire line joining them. We already have the congruence space S, in §1.5. Now an additional feature is needed for its vector space V. With this, translation vectors are comparable for similarity, or being representative of the same direction. Any vector x can be scaled to produce another xt, by the operation of multiplication with a scalar $t \neq 0$. Then the similarity relation $\sim$ between vectors is defined by

$$x \sim y \equiv y = xt \text{ for some } t \neq 0.$$

This relation is reflexive, transitive and symmetric, and so an equivalence relation, the classes of which define *directions*. When vectors are similar, as with $y = xt$, they also have a *proportion*, given by t.

A line can now be characterized by all displacements between its points having the same direction, this direction defining the direction of the line. Hence, taking any two points I, O we can describe all points P on the line, by taking the vector $x = [O, I]$ from O to I, and then $P = O + xt$. The points on the line are now in a correspondence with the scalar parameter t, where with $t = 0$ or 1 we have $P = O$ or I, and other values provide other points of the line.

A linear space L is means for representation of such an affine space S, and its vector space V, in fact where both $S = L$ and $V = L$. Despite these coincidences, a different point of view is present than when L is just a linear space, indicated when L is submitted as representing an affine space. We have another representation under an affine transformation, that preserves the similarity and proportion of similar displacements.

With $a, b \in L$, and $d = b - a$, we have $d \in L$, and $a + d = a$. This shows the accident that allows $S = L$ and $V = L$, and the $+$ between S and V to be the $+$ in L.

For the line M through a, b we require $y - x \sim d$ for $x, y \in M$, in particular $x - a \sim d$ for $x \in M$, that is, $x = a + dt$ for some t. Then M is locus of such x. By taking $s = 1 - t$, we have the line through a, b described by

$$x = as + bt, \quad \text{where } s + t = 1.$$

Now for an *affine subspace* M, we require it to contain the entire line through any pair of its points, and hence that

$$x, y \in M, \ s + t = 1 \ \Rightarrow \ xs + yt \in M.$$

As follows by induction, this condition is equivalent to

$$x, y, \ldots \in M, \ s + t + \cdots = 1 \ \Rightarrow \ xs + yt + \cdots \in M.$$

We may call a linear combination of elements where the coefficients sum to 1 an *affine combination* (so if also they are non-negative we have, in usual terms, a convex combination). Then the characteristic of an affine subspace is that it contain all affine combinations of its elements. Then L and all its linear subspaces are affine subspaces, and any affine subspace has the form $d + V$, where V is a linear subspace, so it is a translation of some linear subspace of L by a displacement $d \in L$. Hence if S is an affine subspace, by taking any of its points $d \in S$, we have $V = M - d$ is a linear space, independent of the particular d, this being the unique vector space associated with the affine space S, so if d is any point of S we have $S = d + V$.

For an affine transformation f we require

$$f((y - x)t) = (fy - fx)t,$$

so similarity and proportion are preserved. Such a transformation has the form

$$fx = d + ax,$$

where $d \in L$, and a is a linear transformation of L. By such a transformation, affine subspaces are mapped onto affine subspaces.

3
Linear Dependence

3.1 Linear dependence

Elements $x_1, \dots, x_p$ are *linearly independent* if no *proper* linear relation between them exists, so

$$x_1 t_1 + \cdots + x_p t_p = o \implies t_1, \dots, t_p = 0.$$

Otherwise they are *linearly dependent*, in which case

$$x_1 t_1 + \cdots + x_p t_p = o, \text{ where some } t_i \neq 0,$$

say $t_1 \neq 0$. Then

$$x_1 = x_2(-t_2/t_1) + \cdots + x_p(-t_p/t_1),$$

so

$$x_1 \in [x_2, \dots, x_p].$$

Also, if we have this, say

$$x_1 = x_2 t_2 + \cdots + x_p t_p,$$

then

$$x_1 1 + x_2(-t_2) + \cdots + x_p(-t_p) = o,$$

so $x_1, \dots, x_p$ have a proper linear relation, showing their linear dependence. Accordingly,

Theorem 1.1 For any given elements, either they are independent, or some one of them lies in the span of the others, and not both.

Also, directly from the definition,

Theorem 1.2 Any subset of independent elements is independent.

Further,

Theorem 1.3 If any elements are independent, but are not when adjoined with a further element, then the further element belongs to their span.

That is, if $x_1, \ldots, x_p$ are independent, but $x_0, x_1, \ldots, x_p$ are not, then $x_0 \in [x_1, \ldots, x_p]$. For, from the dependence,

$$x_0 t_0 + x_1 t_1 + \cdots + x_p t_p = 0,$$

where not all t's $= 0$. But then, from the given independence, it is impossible that $t_0 = 0$.

A *maximal independent subset* of any set of elements is a subset which is independent, but any larger subset containing it is not.

Theorem 1.4 The maximal independent subsets of a given set are identical with the independent subsets with span that includes the entire set.

This is a corollary of the last theorem.

Theorem 1.5 Every finite set of elements in a linear space has a maximal independent subset.

The set should include some non-null element, or there is nothing more to prove. Suppose some r independent elements have been found. If $r = p$, or the remaining elements are already in the span of the r, we already have a maximal independent subset. Otherwise one of the remaining elements not in the span can be adjoined to obtain a larger set of $r + 1$ independent elements. The given set being finite, the process of repeated enlargement must terminate, to provide a maximal independent subset.

For an alternative procedure, scanning the given elements in any order, if any element is in the span of the others, remove it. After one pass, a maximal independent set remains.

These arguments for existence presuppose an independence test, which we do not yet have, so the procedures are not offered constructively. If the elements are given not abstractly but, as practically they may be, by coordinates relative to an independent set of generators, or a base, as they always are when we deal with K^n, then, later, we will have an algorithm for determining a maximal independent subset. But beside whatever else, this algorithm itself provides the now wanted independence test, immediately. Hence, when we do have the test, we do not have to follow the construction procedure now described for a maximal independent subset, using the independence test repeatedly. Rather, having the construction procedure then depend on the independence test, as now, would beg the question of the test, since it is from the procedure that we shall have the test; and having an algorithm that presupposes itself would be a quite unlikely arrangement.

3.2 Replacement

A linear combination y of $x_1, \dots, x_p$ is *dependent on* x_r if it is with a non-zero coefficient for x_r, and otherwise *independent of* x_r. By *replacement* now is meant replacement of an element in a set by a linear combination which is dependent on it. Let

$$y = x_1 t_1 + x_2 t_2 + \cdots + x_p t_p$$

be a combination that depends on x_1, so that $t_1 \neq 0$. Then y can replace x_1 in $x_1, x_2, \dots, x_p$ to produce the set $y, x_2, \dots, x_p$.

Theorem 2.1 Replacement leaves span unaltered and preserves independence.

For any combination y of the x's which depends on x_r, we have

$$y = \sum_{i \neq r} x_i\, t_i + x_r\, t_r, \quad t_r \neq 0,$$

and hence

$$x_r = \sum_{i \neq r} x_i\, (-t_i/t_r) + y\, (1/t_r).$$

Now for any z, z' in the spans M, M' obtained before and after the replacement of x_r by y, we have expressions

$$z = \sum_{i \neq r} x_i\, s_i + x_r\, s_r,$$

and

$$z' = \sum_{i \neq r} x_i\, s_i' + y\, s_r'.$$

But, from the expression for y and its rearrangement for x_r, the points z, z' of M, M' connected by

$$s_i' = s_i - t_i\, s_r/t_r \ (i \neq r), \quad s_r' = s_r/t_r,$$

and inversely,

$$s_i = s_i' + t_i\, s_i' \ (i \neq r), \quad s_r = t_r\, s_r'.$$

are in a one-one correspondence, such that $z = z'$. This shows that $M = M'$, as required for the first part.

Again for points z, z' in this correspondence, independence before and after replacement requires

$$z = o \ \Rightarrow \ s_i = 0 \text{ for all } i,$$

$$z' = o \;\Rightarrow\; s'_i = 0 \text{ for all } i.$$

But the left sides are equivalent, since $z = z'$. Evidently, so also are the right sides equivalent. Hence these linear independence conditions are equivalent. QED

Fundamental propositions of the theory of linear dependence are derived from this *Replacement Theorem*.

A *multiple replacement* is a sequence of such simple replacements, drawn from a given set of combinations.

Corollary (*i*) The same holds for multiple replacements.

The proof is by induction on the number of simple replacements.

Corollary (*ii*)   If $y_1, \dots, y_q$ in the span of $x_1, \dots, x_p$ are independent, then they can replace some q of the x's.

The proof is by induction on q, the case $q = 1$ being covered in the course of the argument. Suppose y_i has replaced x_i, for $i < q$, so now we have

$$y_q = \sum_{i<q} y_i\, t_i + \sum_{i \geq q} x_i\, t_i,$$

where $y_q \neq o$ because of the independence of the y's, and hence $t_i \neq 0$ for some i. Then, again because of the independence, we cannot have $t_i = 0$ for all $i \geq q$. Hence $t_i \neq 0$ for some i, say $i = q$. Then y_q can replace x_q. QED

If we combine this with Corollary (*i*), we have the key theorem of linear algebra associated with Steinitz. This has the consequence that follows now, and it is this that allows us to introduce a notion of dimension in the next chapter.

Corollary (*iii*)   There are at most p independent elements in a linear space with p generators.

For any q independent elements in the span can replace some q of the p generators, so $q \leq p$.

Corollary (*iv*)   Any subspace of a finitely generated linear space is finitely generated.

Consider $M < L$, where L has generators $x_1, \dots, x_p$. If $M = O$, then M has o as its single generator. Otherwise there exists some

$y_1 \neq o$ in M. Then y_1, being in L, is a combination of the x's. Suppose some $q \geq 1$ independent $y_1, \ldots, y_q \in M$ have been found. These being independent combinations of the p generators, $q > p$ is impossible, by Corollary (iii). Either their span is M, in which case we already have a set of generators for M, or there exists some element y_{q+1} in M and not in their span. Then $y_1, \ldots, y_{q+1}$ are independent combinations of the x's. This enlargement of the set of y's, starting with $q = 1$, must terminate for some $q \leq p$, to provide a finite set of generators for M. QED

By the preservation of span, any combinations before a replacement remain combinations after, and so continue as candidates for a possible further replacement. A *maximal replacement* is a multiple replacement with a given set of combinations which cannot be taken further with the confinement to that set. Let $y_1, \ldots, y_q$ be combinations of $x_1, \ldots, x_p$ and suppose there has been a multiple replacement in which $y_1, \ldots, y_r$ have replaced $x_1, \ldots, x_r$. This is a maximal replacement if the further y's are not dependent on any of the remaining x's.

***Corollary* (v)** If $y_1, \ldots, y_q$ in the span of independent $x_1, \ldots, x_p$ are such that $y_1, \ldots, y_r$ have replaced $x_1, \ldots, x_r$ and this multiple replacement of x's by y's is maximal, then these r of the y's are independent and the remaining ones are in their span.

Since replacement preserves independence,

$$y_1, \ldots, y_r, x_{r+1}, \ldots, x_p$$

are independent, and hence, by Theorem 3.3, so are $y_1, \ldots, y_r$. After the replacements,

$$y_j = \sum_{i \leq r} y_i t_{ij} + \sum_{i > r} x_i t_{ij} \ (j > r),$$

and no further replacement is possible, so that $t_{ij} = 0 \ (i, j > r)$, and the $y_j \ (j > r)$ are expressed as combinations of the $y_i \ (i \leq r)$. QED

3.3 Rank

Theorem 3.1 The maximal independent subsets of any finite set all have the same number of elements.

Given two such sets, of p and q elements, since the p are independent and generated by the q we have $p \leq q$, by Theorem 2.1, Corollary (*iii*). Similarly $q \leq p$, and so $p = q$.

The *rank* of any finite set of elements in a linear space is defined by the common number of elements in the maximal independent subsets.

Theorem 3.2 Any set of r independent elements in a set of rank r is a maximal independent subset.

For, taken with a further element, we have $r + 1$ combinations of the r elements in any maximal independent subset, so, by Theorem 2.1, Corollary (*iii*), we have a dependent set. Hence this independent set cannot be enlarged, or is maximal. QED

If p elements have rank r, then $p - r$ is their *nullity*. If $r = p$, they are of *full rank;* this is the condition for their independence. The algorithm for finding a maximal independent subset, and hence the rank, now appears as providing an *independence test*.

3.4 Elementary operations

There are two *elementary operations*, with any set of elements in a linear space.

(I) Multiply one element by a non-zero scalar,
$$(\ldots , x, \ldots) \rightarrow (\ldots , xt, \ldots) \ (t \neq 0).$$
(II) Add one element to another,
$$(\ldots , x, \ldots , y, \ldots) \rightarrow (\ldots , x + y, \ldots , y, \ldots).$$

Theorem 4.1 The span of any elements is unaltered by elementary operations.

For
$$y = x_1 t_1 + x_2 t_2 + \cdots + x_p t_p$$
is equivalent to
$$y = (x_1 s)(s^{-1} t_1) + x_2 t_2 + \cdots + x_p t_p \ (s \neq 0),$$
and to
$$y = (x_1 + x_2) t_1 + x_2(t_2 - t_1) + \cdots + x_p t_p.$$
Hence the span of $x_1, x_2, \ldots , x_p$ is unaltered when x_1 is replaced by $x_1 s \ (s \neq 0)$, or by $x_1 + x_2$.

Theorem 4.2 Elementary operations are reversed by elementary operations.

For the inverse of (I), we have

$$(\dots , (xt) \, t^{-1}, \dots) \rightarrow (\dots , x, \dots) \quad (I),$$

which is again elementary. For the inverse of (II), there is the elementary sequence

$$(\dots , x + y, \dots , y, \dots)$$
$$\rightarrow (\dots , x + y, \dots , -y, \dots) \quad (I)$$
$$\rightarrow (\dots , x + y + (-y), \dots , -y, \dots) \quad (II)$$
$$\rightarrow (\dots , x, \dots , y, \dots) \quad (I).$$

Theorem 4.3 Linear independence is preserved under elementary operations.

In the argument for Theorem 4.1, with $y = o$, the independence of $x_1, \dots , x_p$ requires $t_1, \dots , t_p = 0$. But

$$t_1, t_2 = 0 \iff s^{-1} t_1, \; t_2 = 0 \iff t_1, t_2 - t_1 = 0.$$

Theorem 4.4 Elements may be given any permutation by elementary operations.

Any permutation being a product of transpositions, it is enough to consider a transposition

$$(\dots , x, \dots , y, \dots) \rightarrow (\dots , y, \dots , x, \dots).$$

That this can be accomplished by elementary operations is seen from the sequence

$$(\dots , x, \dots , y, \dots)$$
$$\rightarrow (\dots , x + y, \dots , y, \dots) \quad (II)$$
$$\rightarrow (\dots , x + y, \dots , y - (x + y), \dots) \quad (I, II, I)$$
$$\rightarrow (\dots , y, \dots , x, \dots) \quad (I).$$

4
Dimension

4.1 Base

A *base* for a linear space is any independent set of generators.

Theorem 1.1 Any finitely generated linear space has a base.

Take any maximal independent subset of the generators. All the generators, and hence the entire space, lie in its span, so it is an independent set of generators for the space, or a base. Theorem 3.1.5 provides the existence of a maximal independent subset, and a constructive procedure comes from the Replacement Theorem, Corollary (v).

A base being so identified, arguments about rank are repeated in the following that proceed from the definition of a base.

Theorem 1.2 All the bases in a finitely generated linear space have the same number of elements.

This corresponds to Corollary (iv). Suppose we have two bases, of p and q elements. Since the first set is independent, and the second a set of generators, the first is an independent set lying in the span of the second, so that $p \leq q$, by the Replacement Theorem, Corollary (iii); similarly, $q \leq p$, and hence $p = q$.

The common number of elements in bases defines the *dimension* of the space.

Theorem 1.3 Any n independent elements in an n-dimensional linear space form a base.

With any further element we have a dependent set, by Corollary (iii), so this element is in their span. This reproduces Corollary (vi).

4.2 Dimension

Theorem 2.1 For finite dimensional L, if $M < L$, then

$$\dim M \leq \dim L,$$
$$\dim M = \dim L \;\Rightarrow\; M = L.$$

In other words, the only subspace of L with the same dimension is L itself. Any base for M would constitute a set of n independent elements of L, and so be a base also for L.

Theorem 2.2 Any p independent elements in a finite dimensional linear space can replace some p elements in any base to form a new base.

The replacement can be made without altering the span, or the independence, by the Replacement Theorem, Corollary (ii), so now there are n elements that span the space, or a base.

Theorem 2.3 Every subspace of a finite dimensional linear space has a complement.

Elements of a base $e_1, \ldots, e_n$ for L may be replaced by the elements $f_1, \ldots, f_p$ of a base for a subspace $M < L$, to provide a new base for L, say

$$f_1, \ldots, f_p, e_{p+1}, \ldots, e_n.$$

Then the subspace $N < L$ given by

$$N = [e_{p+1}, \ldots, e_n]$$

is a complement of M.

Theorem 2.4 For finite dimensional $M, N < L$,

$$\dim M \vee N = \dim M + \dim N - \dim M \wedge N.$$

Elements in bases $x_1, \ldots, x_p$ and $y_1, \ldots, y_q$ for M and N can be replaced by the elements $z_1, \ldots, z_r$ of a base for the common subspace $M \wedge N < M, N$ to form new bases for M and N, say

$$z_1, \ldots, z_r, x_{r+1}, \ldots, x_p$$

and

$$z_1, \ldots, z_r, y_{r+1}, \ldots, y_q.$$

Then the elements

$$x_{r+1}, \dots , x_p, y_{r+1}, \dots , y_p, z_1, \dots , z_r$$

span $M \vee N$, and it remains to see that they are independent. Any linear relation between them would show $x = y + z$, where x, y and z are combinations of the x's, y's and z's among them. This identifies an element of M with an element of N, and so with an element of the intersection $M \wedge N$. But these elements are in the intersection if and only if $x = o$ and $y = o$, and then also $z = o$. Now from the independence of the subsets of elements in these combinations, their coefficients must be zero. Therefore all the coefficients in the supposed linear relation are zero, showing the required independence. We now have

$$\dim M \vee N = (p - r) + (q - r) + r = p + q - r.$$

QED

Corollary (i) For finite dimensional L and any $M, N < L$,

$$\dim M \vee N \leq \dim L,$$

with equality if and only if $L = M \oplus N$.

For the equality holds if and only if M, N are independent.

Corollary (ii) For finite dimensional L, and any subspaces

$$M, N, \dots < L,$$

we have

$$\dim (M \vee N \vee \dots) \leq \dim L,$$

with equality if an only if

$$L = M \oplus N \oplus \cdots .$$

The proof is by induction on the number of subspaces.

Corollary (iii) For any finite collection of finite dimensional linear spaces L_i,

$$\dim \bigoplus_i L_i = \sum_i \dim L_i.$$

We have this from the last corollary, by isomorphism of direct sum with join in the case of independent subspaces. Also it comes directly, by taking bases in the spaces and from these forming a base for the direct sum.

4.3 Coordinates

Consider a linear space L of dimension n over K, and let $e_1, \ldots, e_n$ be the elements of a base. Since these elements are generators for L, for any $x \in L$ there exist coefficients $t_1, \ldots, t_n \in K$ such that

$$x = e_1 t_1 + \cdots + e_n t_n.$$

Moreover, since the base elements are independent, such coefficients are unique. For should we have

$$x = e_1 s_1 + \cdots + e_n s_n,$$

it would follow that

$$e_1(s_1 - t_1) + \cdots + e_n(s_n - t_n) = 0,$$

and hence, because of the independence, that

$$s_1 = t_1, \ldots, s_n = t_n.$$

The unique coefficients $t_1, \ldots, t_n$ define the *coordinates* of any $x \in L$ *relative to the base* $e_1, \ldots, e_n$. They are the elements of a vector $t \in K^n$, *the coordinate vector* of x relative to the base.

Any base in L so determines a one-one correspondence between the points of L and K^n, or a coordination of L, where $x \to t$ asserts $x \in L$ has coordinate vector $t \in K^n$, the base being understood.

Now the vector space K^n, with the usual addition and scalar multiplication, is itself a linear space over K; and the coordination is an isomorphism between L and K^n, as linear spaces, the considered mapping $f : L \to K_n$ being one-one between the spaces and such that

$$f(xt) = f(x)t, \quad f(x + y) = f(x) + f(y),$$

where $x, y \in L$ and $t \in K$.

The isomorphism enables any discussion of L to be given in terms having reference to K^n—as with Descartes when he made a correspondence between geometric points and coordinates. Though K^n may appear as an example of an n-dimensional linear space over K, nothing is lost about the abstract L when it is taken that $L = K^n$. On the other hand, a discussion of K^n where it is treated like an abstract space L exposes features for abstract algebra.

With the coordination, we have that $o \to o$, linear independence is preserved, and the image of any base in L is a base in K^n. In particular, the image of the coordinate base $e_1, \ldots, e_n$ in L is the base in K^n provided by the vectors $1_1, \ldots, 1_n$, where 1_r has rth element 1 and all other elements zero, making it the rth column $1_r = 1_{r)}$ of the unit matrix 1 of order n.

For any vector $t \in K^n$ with elements $t_1, \ldots, t_n$ we have

$$t = l_1 t_1 + \cdots + l_n t_n.$$

In this way, the elements of a vector in the space K^n, when this is regarded as a linear space, appear as its coordinates relative to the base $l_1, \ldots, l_n$, this base being distinguished as the *fundamental base* in K^n.

Consider an n-dimensional space L with base $e_1, \ldots, e_n$. Any other n elements are given by

$$f_j = \sum_j e_i s_{ij} \tag{i}$$

where the coefficients s_{ij} are unique, and form a matrix s whose columns $s_{j)}$ are the coordinate vectors of the f's relative to the base. If the f's are independent, they also form a base, so we also have

$$e_k = \sum_j f_j t_{jk} \tag{ii}$$

for a unique matrix t. Substituting for the f's from (i),

$$e_k = \sum_j \left(\sum_i e_i s_{ij} \right) t_{jk} = \sum_i \left(\sum_j e_i s_{ij} \right) t_{jk} = \sum_i e_i r_{ik}$$

where the r's are elements of the matrix product $r = st$. Now from the independence of the e's we have

$$r_{ik} = \delta_{ik},$$

where, for the Kronecker δ,

$$\delta_{kk} = 1, \ \delta_{ik} = 0 \ (i \neq k).$$

Hence $r = l$, the unit matrix, so we have $st = l$, and similarly, $ts = l$, showing that s and t are inverse matrices.

Theorem 3.1 If an element in a space has coordinate vectors x, y relative to two bases and s is the matrix of the first base relative to the second and t that of the second base relative to the first, then

$$x = ty, \ y = sx,$$

and

$$st = l = ts,$$

so s, t are inverse matrices.

For we have

$$\sum_i e_i x_i = \sum_j f_j y_j = \sum_j \left(\sum_i e_i s_{ij} \right) y_j = \sum_i e_i \left(\sum_j s_{ij} y_j \right),$$

so that, since the e's are independent,

$$x_i = \sum_j s_{ij} y_j.$$

4.4 Dual base

Consider an n-dimensional linear space L over K, with any base $e_1, \ldots, e_n$ in L, and the dual space L' of linear forms on L. For any $x \in L$,

$$x = \sum_i e_i \, t_i \, ,$$

for unique t_i, the coordinates of x relative to the base. Now for any $u \in L'$,

$$ux = u \left(\sum_i e_i \, t_i \right) = \sum_i \left((ue_i) \, t_i \right) = \sum_i s_i \, t_i,$$

where $s_i = ue_i$. This shows that any $u \in L'$ is fully determined by the values s_i it gives to the elements e_i of any base in L.

Now with the considered base elements $e_i \in L$, let elements $f_i \in L'$ be defined by

$$f_i \, e_i = 1, \; f_i \, e_j = 0 \; (j \neq i).$$

If

$$u = \sum_i s_i f_i,$$

for any $s_i \in K$, then

$$ue_j = \left(\sum_i s_i f_i \right) e_j = s_j (f_j \, e_j) + \sum_{i \neq j} s_i (f_i \, e_j) = s_j.$$

The f_i therefore are generators for all elements in L'. Also they are independent. For if

$$\sum_i s_i f_i = o,$$

for any $s_i \in K$, then

$$s_i = \left(\sum_j s_j f_j \right) e_i = oe_i = 0,$$

so $s_i = 0$ for all i. Hence the f_i are a base in the dual space L' of L, the dual of the base e_i in L for which it has been determined. As a corollary, L' has dimension n.

We may now consider the dual L'' of L'. With reference to section 2.6, the condition, for all $x \in L$,

$$x \neq o \; \Rightarrow \; ux \neq 0 \text{ for some } u \in L'$$

can be satisfied. For if x has coordinate $t_i \neq o$, we can take $u = f_i$. This makes L'', in any case with a natural homomorphism with a subspace of L, isomorphic with its image. Identifying it with its image, we have $L'' < L$ and, since

$$\dim L'' = \dim L,$$

it follows that $L'' = L$, in other words,

Theorem 4.1 Any finite dimensional linear space is reflexive.

The isomorphism between L and L'' can be represented in terms of dual bases. Let g_i be elements of the base in L'' dual to the base f_i in L'. These correspond to the elements e_i for the first base in L. Then any element $g_i t_i \in L''$ corresponds to the element $\sum_i e_i t_i \in L$.

4.5 Annihilator dimension

Theorem 5.1 For finite dimensional L, and any $M < L$,

$$\dim M + \dim M° = \dim L.$$

With $M < L$, any base $e_1, \ldots, e_m$ for M may be extended to a base $e_1, \ldots, e_n$ for L. Let $f_1, \ldots, f_n$ be the dual base in L', and let R be the subspace of L' spanned by $f_{m+1}, \ldots, f_n$. Now

$$x \in M, \ u \in R \ \Rightarrow \ ux = 0,$$

so that $R < M°$. But any element $u \in L'$ is given by

$$u = s_1 f_1 + s_n u_n$$

for some $s_i \in K$. Then

$$u \in M° \Rightarrow s_i = ue_i = 0 \ (i = 1, \ldots, m) \ \Rightarrow \ u \in R,$$

showing that also $M° < R$, and hence $M° = R$. Now we have

$$\dim M° = \dim R = n - m = \dim L - \dim M.$$

QED

Theorem 5.2 For finite dimensional L, and any $M < L$, $M°° = M$.

By the definitions of $M°$ and $M°°$,

$$x \in M . \Rightarrow . u \in M° \Rightarrow ux = 0 . \Rightarrow . x \in M°°,$$

so that $M < M°°$. But by the last theorem,

$$\dim M = \dim M°°.$$

Hence $M = M°°$. QED

Theorem 5.3 $L = M \oplus N \Rightarrow L' = M° \oplus N°.$

Because of what we have about dimensions, it is enough to show that

$$M° \cap N° = O.$$

Consider any $u \in M° \cap N°$. It will be shown that $u = o$.

We have
$$x \in M, \ y \in N \ \Rightarrow \ ux = 0, \ uy = 0.$$
But, with $L = M \oplus N$,
$$z \in L \ \Rightarrow \ z = x + y \text{ for some } x \in M, \ y \in N.$$
so now we have
$$z \in L \ \Rightarrow \ uz = 0,$$
which implies $u \in L^\circ$. But $L^\circ = O$. Hence $u = o$. QED

In this case there is the natural isomorphism between M° and N', with correspondence between $u \in M^\circ$ and $v \in N'$ such that $uz = vz$.

4.6 Affine coordinates

Beside the coordinate system for an n-dimensional linear space L with reference to a base of n independent elements, another system is familiar, especially from analytic geometry. These are the barycentric coordinates, which have reference to $n + 1$ elements $a_0, a_1, \ldots, a_n$ which are the vertices of a regular n-simplex Δ. These are invariant under translations, and all affine transformations of L, and so have a suitability when L is regarded as representing an affine space $S = L$, with $V = L$ as its vector space. Hence here these will be called *affine coordinates*, and the others can be distinguished as *linear coordinates*.

The affine coordinates relative to Δ express any element $x \in S$ as a unique combination

$$x = a_0 t_0 + a_1 t_1 + \cdots + a_n t_n, \tag{i}$$

where the coefficients which are the coordinates sum to 1,

$$t_0 + t_1 + \cdots + t_n = 1, \tag{ii}$$

Both the general possibility of the representation and its uniqueness depend on Δ being a regular simplex, where the vertices have affine independence, so no vertex lies in the affine subspace spanned by the others. Any translation mapping $x \rightarrow x + d$ of S leaves these coordinates unchanged.

Such coordinates were called barycentric because, at least when the t's are non-negative, since they sum to 1, they represent a distribution on the vertices of the simplex, as of a unit mass, and then x would be the centroid.

For the simplex Δ to be regular, the vectors on the n adjacent edges at any vertex must be linearly independent. For instance, taking the vertex a_0, the vectors

$$a_1 - a_0, \ldots, a_n - a_0$$

should be independent. The similar conditions obtained by taking any other vertex are equivalent to this, and to the affine independence of the vertices, that is, the impossibility of a linear relation between them where the coefficients sum to 1.

Because of the linear independence of the vectors on the edges at a_0, the vector $x - a_0$ has a unique expression as a linear combination of these vectors,

$$x - a_0 = (a_1 - a_0)t_1 + \cdots + (a_n - a_0)t_n.$$

Hence, taking

$$t_0 = 1 - t_1 - \cdots - t_n,$$

we have the expression (i) where the coefficients satisfy (ii).

5
Replacement

5.1 Replacement algorithm

Elements $y_1, \ldots, y_q$, in a linear space over a field K, are given as combinations of elements $x_1, \ldots, x_p$ by the *linear dependence table*

$$
\begin{array}{ccccc}
y_1 & \cdots & y_s & \cdots & y_q \\
x_1 & t_{11} & \cdots & t_{1s} & \cdots & t_{1q} \\
\vdots & \vdots & \cdots & \vdots & \cdots & \vdots \\
x_r & t_{r1} & \cdots & t_{rs} & \cdots & t_{rq} \\
\vdots & \vdots & \cdots & \vdots & \cdots & \vdots \\
x_p & t_{p1} & \cdots & t_{ps} & \cdots & t_{pq}
\end{array}
\tag{i}
$$

which informs that

$$y_j = \sum_i x_i \, t_{ij},$$

where $t_{ij} \in K$.

Provided $t_{rs} \neq 0$, the generator x_r can be *replaced* by the combination y_s, to obtain a new table

$$
\begin{array}{ccccc}
y_1 & \cdots & y_s & \cdots & y_q \\
x_1 & t'_{11} & \cdots & 0 & \cdots & t'_{1q} \\
\vdots & \vdots & \cdots & \vdots & \cdots & \vdots \\
y_s & t'_{r1} & \cdots & 1 & \cdots & t'_{rq} \\
\vdots & \vdots & \cdots & \vdots & \cdots & \vdots \\
x_p & t'_{p1} & \cdots & 0 & \cdots & t'_{pq}
\end{array}
\tag{ii}
$$

where

$$t'_{rj} = t_{rj}/t_{rs}, \quad t'_{ij} = t_{ij} - t_{is}\, t_{rj}/t_{rs} \; (i \neq r).$$

In other words, to produce the table t', row r in table t is divided by t_{rs}, and then such multiples of it are subtracted from the other rows that their elements in column s become 0.

According to the Replacement Theorem, of §3.2, *the span and independence of generators is unaffected by replacement.*

A *multiple replacement*, where $y_1, \ldots, y_r$ replace $x_1, \ldots, x_r$, produces a table

$$
\begin{array}{lcccc}
 & y_1 \;\ldots\; y_r \;\ldots\; y_j \;\ldots & & & \text{(iii)} \\[4pt]
y_1 & 1 \;\ldots\; 0 \;\ldots\; a_{1j} \ldots \\
 & \vdots \;\;\; \vdots \;\ldots\; \vdots \;\ldots\; \vdots \;\ldots \\
y_r & 0 \;\ldots\; 1 \;\ldots\; a_{rj} \ldots \\
 & \vdots \;\;\; \vdots \;\ldots\; \vdots \;\ldots\; \vdots \;\ldots \\
x_i & 0 \;\ldots\; 0 \;\ldots\; a_{ij} \ldots \\
 & \vdots \;\;\; \vdots \;\ldots\; \vdots \;\ldots\; \vdots \;\ldots \\
\end{array}
$$

where $a_{ij} = 0 \; (i,j > r)$ if this is a *maximal replacement*.

The *Replacement Algorithm* proceeds from an initial table to such a termination, possibly restricting replacements to a subset of the y's. This already has had applications, to the Replacement Theorem and consequences. Lastly it was for finding a maximal independent subset of elements given as linear combinations of some independent elements. It has further importance, especially for dealing with systems of linear equations. The replacement operation itself has still further scope, extending into the treatment of systems of linear inequalities, including the topic of linear programming (LP), where its use suggested the application given to it now more broadly.

A review of schematics in the tables (i), (ii) and (iii) will assist future reference. The linear dependence table has the form

$$
\begin{array}{cc}
 & y \\
x & t
\end{array}
$$

t being the matrix of coefficients expressing the y's as combinations of the x's. For the case where x is the fundamental base in K^m, provided by the columns of the unit matrix $1 \in K_m^m$, and y consists in the columns of a matrix $a \in K_n^m$, we simply have $t = a$. This comes from the accident that *any vector has coordinates relative to the fundamental base given by its own elements.* For this case, therefore, the scheme becomes

$$\begin{array}{cc} & a \\ 1 & a \end{array}$$

Now consider the x and y elements partitioned as x', x'' and y', y'' and so, correspondingly, we have the partitioned table

$$\begin{array}{ccc} & y' & y'' \\ x' & p & u \\ x'' & v & d \end{array} \qquad (\text{i})'$$

We suppose the y' are to replace the x', so p will be a square submatrix of t, of order equal to the number of replacements. After the replacements, the table becomes

$$\begin{array}{ccc} & y' & y'' \\ y' & 1 & u^* \\ x'' & o & d^* \end{array} \qquad (\text{ii})'$$

so *if this is a maximal replacement, then $d^* = o$.*

For the present suppose the scheme is for a single replacement, so p is 1×1 and its single element, the pivot element for the replacement, is non-zero, so p^{-1} is defined. Then we have

$$u^* = p^{-1}u, \quad d^* = d - vp^{-1}u.$$

Hence we can represent the replacement as carried out by a matrix premultiplication of the coefficient matrix t, in partitioned form shown by

$$\begin{bmatrix} p^{-1} & o \\ -vp^{-1} & 1 \end{bmatrix} \begin{bmatrix} p & u \\ v & d \end{bmatrix} = \begin{bmatrix} 1 & p^{-1}u \\ o & d-vp^{-1}u \end{bmatrix}$$

A replacement is selected by the choice of a *pivot element*, any non-zero element in the table, lying in the *pivot row* and *pivot column*, which tells which generator element is to be replaced by which combination. With this 'pivot' terminology, we can describe the replacement operation again: *divide the pivot row by the pivot element, and then subtract such multiples of it from the other rows to bring their elements in the pivot column to zero.*

5.2 Matrix rank

An $m \times n$ matrix $a \in K_n^m$ has *row rank* determined as the rank of its m rows

$$a_{(i} \in K_n \ (i = 1, \dots, m),$$

when K_n is regarded as a linear space over K with these as elements. Similarly, the matrix has a *column rank*, having reference to its n columns

$$a_{j)} \in K^m \ (j = 1, \dots , n).$$

It will be concluded that matrix rank defined in these two ways gives the same value. This common value can then be taken without ambiguity to define the *rank* of a matrix.

The replacement operation on a linear dependence table is carried out by a set of elementary operations on the rows of the table. This is apparent from the description in §1. Therefore row rank and row span are unaltered by replacement.

Consider the algorithm by which we would discover the column rank of a matrix $a \in K_n^m$. With reference to §1, we start with the scheme

$$\begin{array}{cc} & a \\ 1 & a \end{array}$$

or in partitioned form

$$\begin{array}{ccc} & a' & a'' \\ 1' & p & u \\ 1'' & v & d \end{array} \tag{i}$$

where a' stands for r independent columns of a, and $1'$ for the fundamental base vectors to be replaced by them. After the replacements we have

$$\begin{array}{ccc} & a' & a'' \\ a' & 1 & u^* \\ 1'' & o & d^* \ (=o) \end{array} \tag{ii}$$

This being a maximal replacement, we have $d^* = o$. The column rank of a is now determined to be r. For, since the original independence of $1'$, $1''$ is preserved by replacement, we have a', $1''$ is independent. Hence the subset a' *is independent.* Also we have all a'' *expressed as linear combinations of the* a', the matrix b^* providing the coefficients, as by the table

$$\begin{array}{cc} & a'' \\ a' & u^* \end{array}$$

This establishes that *the* a' *are a maximal independent set of columns of* a, and hence that *the column rank is* r.

The matrix of the final table is

$$\begin{array}{cc} 1 & u^* \\ o & o \end{array}$$

which has r independent rows and the rest null, so *the row rank is r.* But row rank is unaltered by replacements. Hence this is the row rank of a, now equal to the column rank.

Theorem 2.1 For any matrix, row rank equals column rank.

Here is another argument. Let a have row and column rank r and s, and suppose $a_{j)}$ $(j = 1, \dots , s)$ is a column base. Then

$$a_{k)} = \textstyle\sum_{j \leq s} a_{j)} \, t_{jk}$$

for some t_{jk}, that is

$$a_{ik} = \textstyle\sum_{j \leq s} a_{ij} \, t_{jk} ,$$

so that

$$a_{(i} = \textstyle\sum_{j \leq s} a_{ij} \, t_{(j},$$

showing that $r \leq s$, since no more than s combinations of s vectors can be independent. Similarly $s \leq r$, and hence $r = s$.

The *rank* of any matrix may now be indiscriminately its row rank or column rank. A matrix is of *full rank* if it is of full rank either for its rows or for its columns, so for an $m \times n$-matrix of rank r, full rank requires $r = m$ if $m < n$, and otherwise $r = n$.

Theorem 2.2 Any square matrix has a left inverse if and only if it has a right inverse, and this is if and only if it is of full rank; then these inverses are identical, and unique.

Let b be a right inverse of a, so $ab = 1$, equivalently

$$ab_{j)} = 1_{j)} \ (j = 1, \dots , n),$$

so the columns of 1 are expressed as linear combinations of the columns of a, the coefficients being given by the columns of b. Hence the column rank of 1, which is n, cannot exceed that of a, which is at most n. This shows a is of column rank n. Conversely, if a is of full column rank n, its columns constitute a base in K^n. If b is the matrix whose columns are the coordinate vectors of the columns of 1 relative to this base, we have $ab = 1$, that is, b is a right inverse of a. Hence a has a right inverse if and only if its column rank is n. Similarly it has a left inverse if and only if its row rank is n. But row rank equals column rank. Finally, if b, c are right and left inverses so

$$ab = 1, \quad ca = 1$$

we have

$$c = c1 = c(ab) = (ca)b = 1b = b,$$

and so
$$b = c.$$
Similarly, were b' another right inverse, we would have $b' = c$, and hence $b' = b$; and similarly with the left. QED

A matrix a is *regular* if it has an inverse, a matrix denoted a^{-1} such that
$$aa^{-1} = 1 = a^{-1}a,$$
so it is both a left and right inverse, and unique. Otherwise the matrix is singular.

Corollary A square matrix is regular if and only if it is of full rank.

From the symmetry joining a matrix with its inverse, we have that if b is the inverse of a, then also b is regular, and has a as inverse. Hence, a^{-1} being the inverse of a, we have a is the inverse of a^{-1}, that is,
$$(a^{-1})^{-1} = a.$$
The procedure for finding the rank of a matrix can, with some modification, also find the inverse, if there is one. Instead of with

$$\begin{array}{cc} & a \\ 1 & a \end{array}$$

we start with the augmented table

$$\begin{array}{ccc} & a & 1 \\ 1 & a & 1 \end{array}$$

and follow the same procedure, though restricting the pivot columns to exclude the fundamental base elements, so the pivot elements are in the original part of the table. After a maximal replacement, so restricted, if the matrix is regular we will have a table

$$\begin{array}{ccc} & a & 1 \\ a & 1 & b \end{array}$$

which asserts $1 = ab$, so b is the inverse. Otherwise, the rank is less than n, and the matrix singular. In that case a maximal independent set of columns will have been found, and the rank determined.

5.3 Block replacement

With reference to §4, tables (i)' and (ii)', again consider a partitioned linear dependence table

$$
\begin{array}{ccc}
 & y' & y'' \\
x' & p & u \\
x'' & v & d
\end{array} \qquad \text{(i)}
$$

where the x' and y' elements have the same number, so p is square. For the case where p is a single element, provided this is non-zero so it may become a pivot, we may perform the single replacement of x' by y' to obtain the table

$$
\begin{array}{ccc}
 & y' & y'' \\
y' & 1 & u^* \\
x'' & o & d^*
\end{array} \qquad \text{(ii)}
$$

But now we take p in table (i) to be a square matrix, and consider the possibility of making the block replacement by which the several x' become replaced by the several y', to arrive at a corresponding table of the form (ii). It is to be seen that if p is a regular square matrix, then this is possible.

For simplicity and the sake of argument, though it makes no other difference, suppose all elements are in an n-dimensional space, and elements are referred to by coordinate vectors relative to the same base in this space. We may then understand x' to be a matrix of row order n whose columns are the coordinate vectors of elements in the x' group, and similarly with the other groups. Then table (i) tells that

$$
y' = x'p + x''v,
$$
$$
y'' = x'u + x''d.
$$

Hence if p is regular with inverse p^{-1}, it would follow that

$$
y'p^{-1} = x'pp^{-1} + x''vp^{-1}
$$

and hence that

$$
x' = y'p^{-1} + x''(-vp^{-1}),
$$

so, by substitution, we also have

$$
y'' = y'p^{-1}u + x''(d - vp^{-1}u),
$$

and hence the table (*ii*) with

$$
u^* = p^{-1}u,
$$
$$
d^* = d - vp^{-1}u.
$$

This shows the block replacement, with p as the pivot block, is possible provided p is regular.

To be seen now is that every multiple replacement is equivalent to a single block replacement. In the passage from (i) to (ii), the matrix p becomes specified, and we just have to see that it must be regular. Then taking this as pivot block, we have a corresponding block replacement, equivalent to the multiple replacement. With correspondence so understood, the above may be amplified.

Theorem 3.1 For every multiple replacement, the corresponding single block replacement is feasible, and equivalent to it.

In going from (i) to (ii) by a series of single replacements, all the pivot elements are in the area of p, and p becomes eventually 1, the unit matrix. But the passage from p to 1 is represented, as concerns just the area of p, as a series of elementary operations on the rows, preserving row rank. Hence, since 1 is regular, so is p. QED

An issue that remains is to settle the following.

Theorem 3.2 Any block replacement is expressible as a multiple replacement.

That is, the block replacement can be effected by a series of elementary replacements where generators are replaced one at a time. This can be affirmed. For, with p regular, replacements with pivot elements in the area of p will bring p to 1.

For any matrix, a *critical submatrix* is a square submatrix which is regular, while every square submatrix of greater order that contains it is not.

Theorem 3.3 A matrix is of rank r if and only if it has a critical submatrix of order r.

If a is of rank r then, with reference to tables (i) and (ii) of the last section, and arguments of this section, we have p as a regular square submatrix of a of order r. Also, because matrix rank equals column rank, no columns of a, or of any submatrix, can have rank higher than that of a. Therefore no square submatrix of order greater than r is regular, and hence p is a critical.

For the converse, suppose p in table (i) is a critical submatrix of order r. Then the rank of a is at least r. The replacement with p as pivot block produces table (ii). Were the rank of a greater that r, we would not have $d^* = 0$, and a further replacement would be possible, and we would find a regular square submatrix that enlarges p. But p is critical, so this is impossible. Hence the rank is r. QED

Theorem 3.4 If

$$a = \begin{bmatrix} p & u \\ v & d \end{bmatrix}$$

where p is critical for a, then

$$a = \begin{bmatrix} p \\ v \end{bmatrix} p^{-1} [p \quad u].$$

This is the *critical decomposition* formula for a matrix, obtained in respect to any critical submatrix, and the rows and columns through it. The expression, with the association taken one way, shows all the rows of a as linear combinations of the rows $[p \ b]$ through the critical submatrix, and, taken the other way, does similarly for columns.

With reference to the table (ii) in the last section, with $d^* = o$ we have $d = vp^{-1}u$. QED

5.4 Extended table

When generators in a linear dependence table have been replaced, they are lost, so the process cannot be reversed. This defect can be remedied by including the generators among the generated elements. Instead of with the table

$$\begin{matrix} & y \\ x & a \end{matrix}$$

we start with the *extended table*

$$\begin{matrix} & y & x \\ x & a & 1 \end{matrix}$$

Taking this in the partitioned form

$$\begin{matrix} & y' & y'' & x' & x'' \\ x' & p & u & 1' & o \\ x'' & v & d & o & 1'' \end{matrix}$$

where the y' are to replace the x', so p is the pivot, the table after replacement is

$$\begin{matrix} & y' & y'' & x' & x'' \\ y' & 1' & p^{-1}u & p^{-1} & o \\ x'' & o & d - up^{-1}u & -vp^{-1} & 1'' \end{matrix}$$

Now it is possible to replace generators y' by x', with p^{-1} as pivot, and so return to the original table.

The information in the extended table where we have the unit matrices $1'$, $1''$ is spurious since we can in any case know they occur where they do. If their columns in the table were deleted, they could always be restored for replacement routine purposes. The essential parts of the original and final tables are in the *condensed tables*

$$
\begin{array}{c c c}
 & y' & y'' \\
x' & p & u \\
x'' & v & d
\end{array}
$$

and

$$
\begin{array}{c c c}
 & x' & y'' \\
y' & p^{-1} & p^{-1}u \\
x'' & -vp^{-1} & d - vp^{-1}u
\end{array}
$$

The rule, expressed here, by which we go from the first to the second, when applied to the second, restores the first. This scheme therefore does have the reversibility that was lacking before.

The step from the one table to the other we call a *pivot operation*. This form of operation is associated with A. W. Tucker. It is distinguished from the ordinary replacement operation, which would instead have produced the table

$$
\begin{array}{c c c}
 & y' & y'' \\
y' & 1 & p^{-1}u \\
x'' & o & d - vp^{-1}u
\end{array}
$$

To inform that this operation is in view, the table involved, which has been envisaged as a condensed extended table, will be called a *tableau*, again following Tucker's teminology.

The pivot operation with a tableau is more economical that the replacement operation with an extended linear dependence table. But there is more to it than that. This is where it possesses a sort of symmetry, or duality.

5.5 Condensed table

A matrix $a \in K_n^m$ can be understood to provide the coefficients in a linear dependence table

$$
\begin{array}{c c}
 & y \\
x & a
\end{array}
$$

where the x are elements in a linear space, the generators, and the y are linear combinations of these with coefficients given by a.

Instead now the matrix a is to be associated with relations

$$y = ax$$

between variables given as elements of vectors $x \in K_n$, $y \in K_m$ (not to be confused with the earlier x, y), and similarly, just as well,

$$t = sa,$$

where $s \in K_m$, $t \in K_n$. The pivot operation, which was regarded above as a way of carrying out a replacement, has a significance also for these relations.

Let the matrix a be partitioned as

$$a = \begin{bmatrix} p & u \\ v & d \end{bmatrix}$$

where p is the pivot element, or, equally well, the pivot block. Correspondingly, let x be partitioned into vectors x', x'' and so forth with the other variables. Then we have

$$y' = px' + ux'',$$
$$y'' = vx' + dx''.$$

The independent x variables are distinguished as *base* variables, and the dependent y as *non-base*. Suppose we want to make an exchange of role between the base x' and non-base y'. With p non-zero, if this is a single element, or regular in the case of a block, we have

$$x' = p^{-1}y' + (-p^{-1}u)x'',$$

and then, by substitution for x' in the expression for y'',

$$y'' = vp^{-1}y' + (d - vp^{-1}u)x'.$$

The new coefficient matrix

$$\begin{bmatrix} p^{-1} & -p^{-1}u \\ vp^{-1} & d - vp^{-1}u \end{bmatrix}$$

is identical with that which would be obtained by the pivot operation.

The way we have put it here, there is some apparent difference. In the condensed table operation, a minus is attached to vp^{-1}, whereas with the Tucker operation it is attached to $p^{-1}u$. The correspondence therefore requires a transposition, or dualization, not disturbing what has been said, as clarified further by the next remark.

This further view of the pivot operation as for exchange between base and non-base variables is capable of a double understanding, with reference simultaneously to the coefficient matrix and its transpose. Consider the similar exchange with the other considered relation, $t = sa$, related by transposition to the first. In partitioned form, this is

$$t' = s'p + s''v,$$
$$t'' = s'u + s''d.$$

Then

$$s' = t'p^{-1} + s''(-vp^{-1}),$$
$$t'' = t'(p^{-1}u) + s''(d - vp^{-1}u),$$

so the new coefficient matrix for this case is identical with that obtained for the other. Accordingly, the *same pivot operation* can be understood *simultaneously* as bringing about the exchange between x', y' in the one system, or between u', v' in the other.

The ambiguity in the exchange interpretation has significance for linear programming, in the way Dantzig's Simplex Algorithm has been treated by A. W. Tucker. There it comes out as ambiguity between pivot operations on one or other of a dual pair of LP problems, so an operation on one is simultaneously a similar operation on the other.

6
Linear Equations

6.1 Maximal replacement

Consider a matrix $a \in K_n^m$ and vector $q \in K^m$, and the problem of finding a general solution of the linear equation system

$$ax = q, \qquad \text{(i)}$$

for $x \in K^n$. With a particular solution y, we have

$$a(x - y) = o.$$

Hence the general solution of the system has the form $x = y + z$, where y is a *particular solution*, and z is the general solution of the *reduced system*

$$az = o. \qquad \text{(ii)}$$

The problem of finding a general solution of the system is now broken down into finding a particular solution, together with a general solution of the reduced system. The problem is not broken down in this way when we use the replacement algorithm. Instead, when it terminates, the algorithm immediately provides a general solution, or shows there is no solution.

With

$$
\begin{array}{ccc}
 & a & q \\
l & a & q
\end{array}
$$

as the initial linear dependence table, the algorithm consists in performing a maximal replacement with pivots restricted to the columns of a, that is, excluding the q column. We may partition this table in the form

$$
\begin{array}{ccc}
 & a' & a'' & q \\
l' & p & u & q' \\
l'' & v & d & q''
\end{array}
$$

where, in the *maximal replacement* to follow, shown by d having become o so no further replacement is possible, a' is to replace l', p being the associated pivot block. Then for the final table we have

$$\begin{array}{ccc} & a' & a'' & q \\ a' & 1 & h & x' \\ 1'' & o & o & x'' \end{array}$$

All is available now for knowing whether there is a solution, and if there is one, then having the general solution. From §4.3, it will be known that p must be regular, and

$$h = p^{-1}u, \ x' = p^{-1}q', \ x'' = q'' - vp^{-1}q', \ d = vp^{-1}u,$$

but now we do not depend at all on this.

The condition for the *existence of a solution* is that $x'' = o$. Given this, we have

$$q = a'x' = [a' \ a''] \begin{bmatrix} x' \\ o \end{bmatrix} = ay,$$

where

$$y = \begin{bmatrix} x' \\ o \end{bmatrix}.$$

Therefore y is a *particular solution*.

We also have

$$a'' = a'h,$$

that is,

$$[a' \ a''] \begin{bmatrix} h \\ -1 \end{bmatrix} = o,$$

or

$$ak = o,$$

where

$$k = \begin{bmatrix} h \\ -1 \end{bmatrix}.$$

Then $z = kt$, for any vector $t \in K^{\nu}$, where ν is the column nullity of a, is the general solution of the reduced system, the ν columns of k being a base for the linear subspace of K^n which describes all solutions.

This description requires substantiation on some points, and there can be comment for special cases. About the particular solution, we have a' as a column base for a, and $a', 1''$ as a base for K^m. Then x', x'' provide the unique coordinates for q relative to this base in K^m, and $x'' = o$ is the condition for q to be in the column space of a, that is, *for a solution to exist*. Then we have a particular solution, as stated.

For the reduced system, the columns of k are independent, and provide a set of solutions. It remains to see that they are generators, and so a base for the solution space. We have the rows of $[l \ h]$ as a base for the row space of a. Hence the reduced system solutions are identical with the solutions

$$z = \begin{bmatrix} z' \\ z'' \end{bmatrix}$$

of

$$[l \ h] \begin{bmatrix} z' \\ z'' \end{bmatrix} = o,$$

that is,

$$z' = -hz'',$$

so that

$$z = \begin{bmatrix} -h \\ l \end{bmatrix} z'' = -kz'' = kt,$$

if we take $t = -z''$. This confirms that *the solutions obtained include all reduced system solutions.*

Consider various special possibilities for the final table:

1.

$$\begin{array}{cc} a & q \\ a \ l & x \end{array}$$

This is for when a is a regular square matrix. Whatever q, there is a unique solution, which appears in the table.

2.

$$\begin{array}{ccc} a' & a'' & q \\ a' \quad l & h & x' \end{array}$$

In this case $n > m$, rank is $r = m$, and column nullity, which determines the dimension of the solution space, is $n - r > 0$. For all q, there is an infinity of solutions.

3.

$$\begin{array}{ccc} a & q \\ a \quad l & x' \\ l'' \quad o & x'' \end{array}$$

This case requires $m > n$, $r = n$. A solution exists provided $x'' = o$, and then it is unique.

For an $m \times n$ matrix of rank r, the *column nullity* is $n - r$. The following came out of the above discussion.

Theorem 1.1 For any matrix a, the dimension of the solution space of the system $ax = o$ is equal to the column nullity of a.

The next theorem can also be seen with reference to the foregoing.

Theorem 1.2 For any matrix $a \in K_n^m$ and vector $q \in K^m$,

> *either*
> $$ax = q \quad \text{for some } x \in K^n,$$
> *or*
> $$wa = o, \ wq = 1 \quad \text{for some } w \in K_m,$$
> *and not both.*

Given the first possibility, so that $ax = q$ for some x, if $wa = 0$, then
$$wq = w(ax) = (wa)x = ox = 0,$$
so the second possibility is denied, and hence it is impossible to have both.

For the rest, this is evident from the above algorithm, when it is considered, as in §4.1, that the final table may be obtained from the initial one by premultiplication with a regular matrix. If w is any row of this matrix corresponding to a non-zero element of x'', the existence of which denies the first possibility, then we have $wa = o$, $wq \neq 0$. It appears moreover that the rows of $-vp^{-1}$ span such w.

Alternatively, though amounting to the same, $wa = o$ is equivalent to
$$w'p + w''v = o,$$
and hence requires
$$w' = -w''vp^{-1}.$$
with arbitrary w''. But
$$x'' = q'' - vp^{-1}q',$$
and so
$$wq = w'q' + w''q'' = -w''vp^{-1}q' + w''q'' = w''x''.$$

Hence, with denial of the solubility of $ax = q$, so $x'' \neq o$, the arbitrary w'' can be chosen to make $w''x'' = 1$. Then we have $wa = o$ and $wq = 1$, as required.

For another proof (the most usual one), $[a' \ q]$ is of rank $r + 1$ if there is no solution. Then every vector of order $r + 1$ is a linear combination of the rows, in particular $[o, \ 1]$. Hence, for some w,

$$w[a'\ q] = [o\ 1],$$

that is,

$$wa' = o, \quad wq = 1.$$

But $a'' = a'h$, so also $wa'' = o$, and hence $wa = o$.

6.2 Gaussian elimination

The elimination method for solving simultaneous equations (the high-school method) goes like this:

ELIMINATION METHOD *If you can, take one equation where there is a non-zero coefficient for a variable, solve for that variable in terms of the others and substitute in the other equations, to obtain a system in the other variables, where the one variable has been* eliminated, *and repeat with this, to termination. One possible termination, with the conclusion there is no solution, is where there is an equation with coefficients all zero, but a non-zero constant. Otherwise, there comes a point where remaining equations, if any, put no restrictions on remaining variables, if any, since both coefficients and constants are zero. Such remaining variables may have arbitrary values, and then the* back-equations *used earlier for the substitutions, taken now in reverse order, successively determine all the other variables in terms of these, if any, or uniquely if there are none.*

Suppose that the system to be considered is

$$ax = q. \tag{i}$$

If $a = o$, the existence of a solution requires also $q = o$, and then x is unrestricted. Hence suppose $a \neq o$, so a has some element $p \neq 0$ which, if necessary after a permutation of rows and columns, can be taken to be leading. Then the system has the partitioned representation

$$\begin{bmatrix} p & u \\ v & d \end{bmatrix} \begin{bmatrix} y \\ z \end{bmatrix} = \begin{bmatrix} r \\ s \end{bmatrix}, \tag{1}$$

that is,

$$py + uz = r, \tag{1.1}$$

$$vy + dz = s. \tag{1.2}$$

Then (1.1) is equivalent to

$$y = (-p^{-1}u)z + p^{-1}r, \tag{1.1'}$$

and by substitution of this expression for y in (ii) we have the system

$$(d - vp^{-1}u)z = s - vp^{-1}r \qquad (1.2)$$

where y has been *eliminated*. With any solution z of system $(1.2)'$, and then y determined from the *back equation* $(1.1)'$, we have any solution of the system (1). Now $(1.2)'$ is the system

$$d^*z = s^*, \qquad \text{(i)}^*$$

where

$$d^* = d - vp^{-1}u, \quad s^* = s - vp^{-1}r.$$

If the previous system (1) had only one equation, this would not be present to put a restriction on z. In that case z can be taken to be arbitrary, and to determine y. Otherwise this system, with one less variable, can be treated like the previous one, and so forth.

In this process, the original system has been replaced by an equivalent one, with the same solutions and the advantage of presenting an immediate way to compute the solutions. The main step can be represented as replacing $ax = q$, given by (i), by

$$tax = tq, \qquad \text{(ii)}$$

where the matrix t is given in partitioned form by

$$t = \begin{bmatrix} p^{-1} & o \\ -vp^{-1} & 1 \end{bmatrix}.$$

Thus,

$$ta = \begin{bmatrix} 1 & p^{-1}u \\ o & d-vp^{-1}u \end{bmatrix}, \quad t = \begin{bmatrix} p^{-1}r \\ s-vp^{-1}r \end{bmatrix}.$$

Put this way, we see the step exactly corresponds to the replacement operation, described in §1. Its next repetition would be to multiply system (ii) by a matrix of the form

$$\begin{bmatrix} 1 & o \\ o & t^* \end{bmatrix},$$

where t^* will serve for (i)^* the same way that t served for (i).

Here the first described scheme for solving equations, where there are replacement operations in a linear dependence table, appears as a particular organization the Gaussian method.

6.3 Rank reduction

Theorem 3.1 For any matrix a, if $a_{rs} \neq 0$, and

$$b = a - a_{s)} a_{rs}^{-1} a_{(r},$$

then

$$at = 0 \Rightarrow bt = 0.$$

Since $a_{(r}t = 0$ is already asserted by $at = o$, the proof is immediate. In other words, linear relations between columns of a are inherited by corresponding columns of b, with exactly the same coefficients; obviously, the same holds for rows.

Since $a_{s)} \neq o$, we can consider a set σ of columns that join with this to form a column base for a. They are $k - 1$ in number, k being the column rank of a.

The corresponding columns of b are independent. For

$$b_{\sigma)} = a_{\sigma)} - a_{s)}\, a_{rs}^{-1}\, a_{r\sigma},$$

so did we have $b_{\sigma)}t = o$, for any $t \neq o$, from here we would have a proper linear relation between elements in a column base for a, which is impossible.

Also, they span the columns of b. For any column x of a has a unique expression

$$x = a_{\sigma)}t + a_{s)}\theta,$$

as a combination of the base columns, and then, by the Theorem, the corresponding column of b is

$$y = b_{\sigma)}t + b_{s)}\theta = b_{\sigma)}t,$$

since $b_{s)} = o$.

We now have $k - 1$ elements making a base, showing this to be the column rank of b. The same holds just as well for rows, whence

Corollary (i) The step from a to b reduces the row and column ranks by 1.

Because of this property, we call this the *rank reduction step*, applied to any matrix a, with any element $a_{rs} \neq 0$ as *pivot*.

By repeating this step, at the kth the column rank will have been brought to zero, and the matrix annihilated, in which case also the row rank is zero. Hence the row rank, which is subject to the same reductions simultaneously, must also have been k originally. This proves

Corollary (ii) For any matrix, the row rank and column rank are equal.

In the main part the following was seen above, and then the remainder is obvious.

Corollary (iii)   In the rank reduction step from a to b, any set of columns of a that, with inclusion of the pivot column, are independent, have corresponding to them, now with exclusion of the pivot column, an independent set of columns of b; and *vice versa*; and the same for rows.

A *rank reduction sequence* is any sequence of rank reduction steps, and one terminating in annihilation of the matrix defines a *rank reduction process*. The number of steps in such a process determines the *rank* of the matrix, in which, its row and column ranks coincide. A *critical submatrix* is a regular square submatrix of maximum order.

Corollary (iv)   The indices of the rows and columns from which the pivots have been drawn in a rank reduction process identify maximal independent subsets of rows and columns of the original matrix, intersecting in a critical submatrix.

This follows from the last corollary. From the discussion below, it is evident that the determinant of the critical submatrix is given by the product of the pivot elements.

The formula

$$a = a_{j)}\, a_{ij}^{-1}\, a_{(i},$$

involving any critical submatrix a_{ij}, on some sets i, j of rows and columns—which Tucker mentioned had still to find its way into print—has a resemblance and connection with the above dealings with rank reduction, produced in response to his further remark.

With the above notation, a_{ij} is the submatrix from rows and columns with indices in sets i and j, and $a_{(i}$, $a_{j)}$ are the submatrices formed from those rows and columns. We arrive at the formula in § 5.3 after a maximal replacement, though originally it was obtained by a use of determinants. It has an application in §9.1 for an especially simple proof of properties of symmetric matrices.

Tucker proposed the following: with any element $a_{ij} \neq 0$, subtract

$$a_{j)}\, a_{ij}^{-1}\, a_{(i}$$

from a, and repeat until a is reduced to zero. One would think the main point settled when it is known, as here in Corollary (ii), that the rank of a is reduced by 1 at each step. We will approach this now the way, which had occurred to me first, using determinants, though we come to those only in the next chapter. My second proof involved an appeal to properties of elementary operations. Tucker objected to going this way, and my remedy is in the above discussion.

If i, j are the sets of rows and columns involved in the reduction process, then $a_{j)}$ would be a column base for a, containing a critical submatrix a_{ij}. Then

$$a_{(i} = a_{ij} h,$$

for some h. Then

$$h = a_{ij}^{-1} a_{(i},$$

and hence, again, the formula above.

Consider

$$a = \begin{bmatrix} p & u \\ v & d \end{bmatrix}$$

where $p \neq 0$. The algorithm step subtracts

$$\begin{bmatrix} p \\ v \end{bmatrix} p^{-1} [p \quad u] = \begin{bmatrix} p & u \\ v & vp^{-1}u \end{bmatrix}$$

from a, to leave

$$\begin{bmatrix} 0 & o \\ o & d - vp^{-1}u \end{bmatrix}.$$

It has to be shown that, if a is of rank r, then

$$A = d - vp^{-1}u$$

is of rank $r - 1$.

Since a must have a critical submatrix containing p, suppose it is leading, and so given by

$$a^* = \begin{bmatrix} p & u^* \\ v^* & d^* \end{bmatrix}$$

with truncations after the first r rows and columns. To complete the argument, it will now be shown that

$$A^* = d^* - v^* p^{-1} u^*,$$

of order $r - 1$, is a critical submatrix of A.

By Schur's identity,

$$|a^*| = p |A^*|,$$

so that, since a^* is regular, A^* must be regular. Also, any submatrix of a containing a^* is singular. Then the same argument shows that the corresponding submatrix of A containing A^* is singular. QED

This further method of discovering the rank of a matrix, together with row and column bases, has complete symmetry in regard to rows and columns, as fits the identity between row and column rank. This is as with the Tucker form of pivot operation for Dantzig's Simplex Algorithm, where there also is a symmetry, which in that case fits LP Duality.

The observation on rank reduction on which the method depends can be put more generally. If a_{ij} is a regular $s \times s$ submatrix of a, through the sets i,j of rows and columns, then subtraction of

$$a_{j)} \, a_{ij}^{-1} \, a_{(i}$$

from a reduces the rank by s, that is,

$$b = a - a_{j)} \, a_{ij}^{-1} \, a_{(i}$$

is of rank $r - s$. The proof is exactly like the proof for $s = 1$. In other words, if p above, instead of being a single elements non-zero element, is a regular $s \times s$ submatrix, then A^* has rank $r - s$. In particular, if $s = r$, then $A^* = O$, and we have our original formula again.

This approach, with rank reduction steps, can be elaborated for solving equations and matrix inversion, as in the last among the programs listed at the end of this book. It also gives an efficient test for a quadratic form, associated with a symmetric matrix, to be definite, or semidefinite. With a symmetric matrix, the reductions can all be based on diagonal elements of the reduced matrices, and the sequence of these must be positive. There is a similar test when there are linear constraints, involving a bordered matrix; see Theorem 9.6.2.

The following has reference to the programs 1 and 4, listed at the end of the book. When matrices

$$P_{rs} = A_{s)} \, A_{rs}^{-1} \, A_{(r}$$

are subtracted repeatedly from A, starting with $A = a$, and with ρ, σ as the sets of rows and columns that have acted so far, the sum of these reductions may be accumulated in a matrix B. Initially, when $A = a$, we have $B = 0$. Also, finally we have $A = O$, and then $B = a$.

At each stage the columns of A are linear combinations of the columns of a, so that $A = aU$, for some square U. Initially $U = I$, and finally $U = O$.

Also, the columns of B are linear combinations of the columns of a with indices in the set σ, so $B = aC$ for some C with non-null columns only in the set σ, so that, equivalently, $B = a_{\sigma)}C_{(\sigma}$. Initially, $C = O$, and finally we have $a = a_{\sigma)}c$ where $c = C_{(\sigma}$ and σ is a column base for a.

When A is modified to

$$A^* = A - P_{rs},$$

the sets ρ, σ become enlarged to ρ^*, σ^* by adjunctions of the elements r, s. Then B becomes modified to

$$B^* = B + P_{rs},$$

and U, C to some U^*, C^*. We are able directly to modify C with reference to A, but not so directly with reference to a. However, we are able to make the reference to a by means of U. By keeping track of U we are therefore able to keep track of C, and hence find the final c.

The scheme set out so far can serve again for finding rank, row and columns bases through a critical submatrix, and the determinant. This is as with the earlier program, §2, which is retained for its simplicity. But it also provides coordinates for non-base columns with reference to the base. Hence it serves also for finding a solution base for the homogeneous equation system $ax = o$.

By adjoining further columns to a, but without extending the range of the reduction elements, we have the solution of non-homogeneous equations, and matrix inversion. For instance, with $[a \vdots I]$, obtained by adjunction of a unit matrix, the corresponding part of the coefficient matrix c, on termination when A becomes null, provides the inverse of a, if this exists, and otherwise the inverse of a critical submatrix. Similarly, by adjoining any column q, a basic solution for $ax = q$ is obtained, or it is established that no solution exists.

6.4 Echelon form

By repeating the operation, first described early in the last section for $a_1, \ldots, a_n$, now with $b_2, \ldots, b_n$, and so forth as far as possible, at the termination, because of rank reduction necessarily at the kth step, there will be a set of elements

$$c_1, \ldots, c_k, o, \ldots, o$$

with the original independence and span, and the *echelon* property

$$c_{ij} = 0 \; (j > i).$$

In other words, allowing for permutations of rows and columns, the original matrix a has been transformed into a matrix of the *echelon* form

$$\begin{bmatrix} E & O \\ * & O \end{bmatrix}$$

where E is a square matrix with elements above the diagonal all zero. By this process with rank reduction, we have an algorithm for discovering the rank of $a_1, \ldots, a_n$ and, by keeping track of columns when they are permuted, identifying a maximal independent subset.

The Gaussian elimination method for solving a system of equations amounts to performing this process with the rows of the matrix of the system.

6.5 Pivotal condensation

The determinant, and the matrix inverse, have explicit algebraical formulae. However, these are not useful for computation, for which there are better procedures. For the determinant, first there is the *pivotal condensation* method, that comes directly out of the Gaussian elimination method. In §2, the transition from one system to the other, (i) to (ii), is carried out by elementary operations, whose effect on the determinant is a factor p^{-1}. We see this from the Weierstrass characteristic properties for the determinant. Alternatively, we may use the multiplication theorem and Laplace expansions, so we know $|t| = p^{-1}$, and then have

$$p^{-1}|a| = |d^*|.$$

This shows the principle of pivotal condensation, by which the evaluation of a determinant of order n is reduced to the evaluation of one of order $n - 1$. Hence, in the case of a regular $n \times n$ matrix a, as the elimination method proceeds with a series of pivots $p_1, p_2, \ldots,$ in the end we have a 1, and then

$$|a| = p_1 \cdots p_n,$$

or, in case a is singular so in the end there is a 0, then

$$|a| = 0.$$

We have already connected the Gaussian elimination step with the replacement operation in a linear dependence table, so there is some repetition in the following. But we will have a way of putting the matter again and taking it further.

Theorem 5.1 If b is the matrix obtained from a square matrix a by a replacement operation with pivot element p, then

$$|b| = |a|/p.$$

Any pivot element being non-zero, this comes from the description of the replacement operation in terms of elementary operations, given in §4.1, and the effects of elementary operations on the determinant shown by the Weierstrass properties (I) and (II) and immediate consequences.

Corollary (i) If b is the matrix obtained from a square matrix a by a multiple replacement with pivot elements $p_1, \ldots, p_s$ then

$$|b| = |a|/p_1 \ldots p_s.$$

This follows by induction.

Corollary (ii) If a square matrix a is reduced to the unit matrix I by replacements with pivot elements $p_1, \ldots, p_n$ then

$$|a| = p_1 \ldots p_n,$$

and if no such reduction is possible then $|a| = 0$.

For then

$$1 = |I| = |a|/p_1 \ldots p_n.$$

Otherwise, a maximal replacement will determine the rank as $r \cdots n$ where r is the number of replacements, so the columns of a are dependent, and $|a| = 0$.

Consider a multiple replacement with pivot elements $p_1, \ldots, p_s$ which from a table

$$\begin{matrix} p & u \\ v & d \end{matrix}$$

for a symmetrically partitioned square matrix a produces a table

$$\begin{matrix} I & u^* \\ o & d^* \end{matrix}$$

Then p is regular, and

$$|p| = p_1 \ldots p_s.$$

Also
$$d^* = d - vp^{-1}u,$$
and
$$\begin{vmatrix} 1 & u^* \\ o & d^* \end{vmatrix} = |a|/p_1 \cdots p_s = |a|/|p|.$$

Further replacements may establish that $d^* = 0$, and also that $a = 0$. Otherwise there will be $n - s$ further replacements, with pivots $p_{s+1}, \ldots, p_n$ producing a final table
$$\begin{matrix} 1 & o \\ o & 1 \end{matrix}.$$

Then it will be established that
$$|d^*| = p_{s+1} \cdots p_n,$$
and also that
$$|a| = p_1 \cdots p_n,$$
so now
$$|a|/|p| = p_{s+1} \cdots p_n = |d^*|.$$

In this way we obtain

Corollary (iii) $\begin{vmatrix} p & u \\ v & d \end{vmatrix} = |p||d - vp^{-1}u|,$

for a symmetrically partitioned determinant, where p is regular.

This is Schur's identity.

We have the row-column symmetry of the determinant, where
$$|a| = |a'|,$$
while here there has been favour for columns. A remedy is in the rank reduction procedure of §3. When a is reduced to o after r steps, with pivots $p_1, \ldots, p_r$ from row and column sets i, j providing a critical submatrix $p = a_{ij}$, we have that
$$|p| = p_1 \cdots p_r,$$
and in particular when a is a regular square matrix, that
$$|a| = p_1 \cdots p_n.$$

6.6 Tucker's scheme

The rows and columns of a Tucker tableau, essentially, just have symbolic *labels*, and with a pivot operation those in the pivot row and column are swapped, while other labels are unchanged.

For the operation on the table, we have

$$
\begin{array}{cccccc}
x & * & & y & & * \\
y & \alpha & \beta & \rightarrow & x & \alpha^{-1} & -\alpha^{-1}\beta \\
* & \gamma & \delta & & * & \gamma\alpha^{-1} & \delta-\gamma\alpha^{-1}\beta
\end{array}
$$

where α is the pivot, and the others are typical other elements, in its row or column, or neither. This can be read just as well in matrix fashion when a partition of the table is represented, with the component α as pivot element, or again just as well, as a pivot block. The x and y show labels that are swapped, while the *-labels are unchanged.

As applied to a matrix a as tableau, with r,s as the pivot row and column, and $i \neq r$, $j \neq s$, the new tableau which results has elements

$$b_{rs} = a_{rs}^{-1}, \qquad b_{rj} = -a_{rs}^{-1} a_{rj},$$
$$b_{is} = a_{is} a_{rs}^{-1}, \qquad b_{ij} = a_{ij} - a_{is} a_{rs}^{-1} a_{rj}.$$

Alternatively r,s may be sets of rows and columns for pivot operations with i,j as their complements. For a partitioned matrix

$$a = \begin{bmatrix} p & u \\ v & d \end{bmatrix}$$

with $p = a_{rs}$ now as pivot block, the result of both the multiple pivot operation and the corresponding single block operation to which it is equivalent is

$$b = \begin{bmatrix} p^{-1} & -p^{-1}u \\ vp^{-1} & d - vp^{-1}u \end{bmatrix}$$

For a *maximal pivot*, after which no additional row and column label swap can be made, we have

$$d - vp^{-1}u = o.$$

At this point various things can be read from the tableau. Firstly, the maximal pivot block p is identical with a critical submatrix of a. Hence we have the rank of a, together with row and column bases. We also have p^{-1}, and so the critical decomposition

$$a = \begin{bmatrix} p \\ v \end{bmatrix} p^{-1} [p \quad u]$$

and if the pivots were $p_1, \ldots, p_r$ then we have the determinant

$$|p| = p_1 \cdots p_r.$$

Also, the columns and rows of

$$\begin{bmatrix} p^{-1}u \\ -1 \end{bmatrix}, \quad [vp^{-1} \quad -1]$$

are bases for the solution spaces of

$$wa = o, \quad ax = o.$$

By adding a row p or column q, or both, to the tableau, but confining the pivots to the area of a, one also finds a particular solution to non-homogeneous equations,

$$wa = p, \quad ax = q.$$

For the case where a is a regular square matrix, the maximal pivot is a complete one producing swaps between all row and column labels. In that case, $p = a$, and the inverse a^{-1} has been found. This is a most efficient way of computing the inverse, with improvement on the maximal replacement method. A computer program is among those at the end of this book.

This scheme, in the way shown by Tucker, is peculiarly suitable for dealing with a dual pair of LP problems

$$\max px : ax \le q, x \ge 0,$$

$$\min up : uq \ge p, u \ge o.$$

describe by a tableau

$$T = \begin{bmatrix} 0 & -p \\ -q & a \end{bmatrix}.$$

Replacing T by $-T'$, we have the same pair of problems, though each becomes stated in the form of the other. Solutions to both problems are obtained simultaneously on termination of the algorithm.[8]

A. W. Tucker's 1960 paper

"... deals with an elementary equivalence relation on matrices, called *combinatorial equivalence* because each equivalence class contains just a finite number of matrices. It stems from an attempt to study for a

---

[8] An account together with a BASIC program is in my 1987 book, Ch. V.3.

general field, rather than an ordered field, the linear algebraic structure underlying the "simplex method" of G. B. Dantzig, so remarkably effective in Linear Programming. However, this structure is outlined here by itself as a topic that seems to have wide applicability and interest."[9]

The topic introduced in the LP connection bears on structure of the Gaussian method. The following comes from a reading of the paper.

The matrices of order $m \times n$ with elements in a field K are to be denoted K_n^m, with $K^m = K_1^m$, $K_n = K_n^1$ for the column and row vectors of order m and n.

A matrix $a \in K_n^m$ is associated with the matrix $A \in K_{m+n}^m$ given by $A = [1\ a]$, obtained by adjoining the unit matrix $1 \in K_m^m$; and b is similarly associated with $B = [1\ b]$.

Let A_π denote the matrix A after its columns have received a permutation π. Application to the unit matrix 1 produces the *permutation matrix* 1_π, with transpose $1'_\pi$ such that

$$1_\pi 1'_\pi = 11 = 1,$$

and so identical with the inverse,

$$1'_\pi = 1_\pi^{-1} = 1_{\pi^{-1}}.$$

Then

$$A_\pi = A 1_\pi.$$

Hence denoting $P = 1_\pi$, we have

$$P^{-1} = 1_{\pi^{-1}} = P',$$

and

$$A_\pi = AP.$$

Also let X_π denote a vector $X \in K^{m+n}$ after its elements have had the permutation, and so, since the elements describe its rows,

$$X_\pi = 1'_\pi X = 1_\pi^{-1} X,$$

and

$$AX = A_\pi X_\pi.$$

Since

$$\mathbb{A} = \begin{bmatrix} -a \\ 1 \end{bmatrix} \in K_n^{m+n},$$

of rank n, is such that

$$A\mathbb{A} = [1\ a] \begin{bmatrix} -a \\ 1 \end{bmatrix} = o,$$

---

[9] "A Combinatorial Equivalence of Matrices", *Proceedings of Symposia in Applied Mathematics*, Volume X, Combinatorial Analysis, 129-40; American Mathematical Society.

and since A is of rank m, it follows that the columns of $\mathbb{A}$ are a base for the solution space $\mathcal{A}$ of

$$AX = o \ (X \in K^{m+n}).$$

The solution space of $A_\pi = AI_\pi$ is

$$\mathcal{A}_\pi = \{I_\pi^{-1}X : AX = o\},$$

so that

$$\mathcal{A}_\pi = \{X_\pi : X \in \mathcal{A}\} = \{I_\pi^{-1}X : X \in \mathcal{A}\} = I_\pi^{-1}\mathcal{A}.$$

A further preliminary observation is that identity of the solution spaces $\mathcal{A}, \mathcal{B}$ associated with any a, b is equivalent to $A = gB$, that is

$$[I \ a] = g[I \ b],$$

for some regular g, which implies $g = I$, and hence $a = b$; alternatively, it is equivalent to $\mathbb{A} = \mathbb{B}h$, that is,

$$\begin{bmatrix} -a \\ I \end{bmatrix} = \begin{bmatrix} -b \\ I \end{bmatrix} h,$$

for some regular h, with the same conclusion. It is also equivalent to

$$A\mathbb{B} = [I \ a]\begin{bmatrix} -b \\ I \end{bmatrix} = o,$$

again with the same conclusion. Such arguments provide more of substance when A is replaced by A_π.

Tucker defines a, b to be *equivalent to within a permutation* π if
$$\mathcal{A}_\pi = \mathcal{B},$$

and to be *combinatorially equivalent* if they are so equivalent for some permutation, which relation is stated $a :: b$. Hence with the relation given by

$$a : \pi : b \equiv \mathcal{A}_\pi = \mathcal{B},$$

there is the definition

$$a :: b \equiv a : \pi : b \text{ for some } \pi.$$

It has been noted that

$$a : I : b \iff a = b,$$

or this relation of a, b within the identity permutation implies their identity. In any case as to be seen, $::$ is an equivalence relation, the classes of which are all finite.

Theorem 6.1 $a : I : a,$

$$a : \pi : b \implies b : \pi^{-1} : a,$$

$$a:\rho:b:\sigma:c \;\Rightarrow\; a:\sigma\rho:c.$$

These are ready consequences of the equivalence of the relation $a:\pi:b$ to the condition

$$[l\,a]\,l_\pi = g\,[l\,b\,]$$

holding for some regular g.

Corollary $a::a,$

$$a::b \;\Rightarrow\; b::a,$$

$$a::b::c \;\Rightarrow\; a::c;$$

so ::, being reflexive, symmetric and transitive, is an equivalence relation.

Theorem 6.2 $a::b$ if and only if there exist G, H with m, n columns forming complementary subsets, in any order, of the columns of $[l\ a]$, such that

$$Gb = H. \tag{i}$$

A matrix pair G, H which constitutes such a column partition of $[l\ a]$ determines, and is determined by, a permutation π, from the relation

$$[l\ a]\,l_\pi = [G\,H]. \tag{ii}$$

If $Gb = H$ for such a pair, then G must be regular, since its columns then form a column base for $[l\ a]$, which is of rank m.

For $a::b$, it is necessary and sufficient that

$$[l\ a]\,l_\pi = g\,[l\ b\,],$$

for some π, and regular g, that is

$$[G\,H] = g\,[l\ b\,],$$

equivalently, $G = g, H = gb$, that is, $H = Gb$, and also conversely, since the G in this conclusion must be regular.

Theorem 6.3 The classes of :: contain at most $(m + n)!$ members.

For there are at most $(m + n)!$ cases of (ii), and a subset of these where G is regular, by (i) to determine $b = G^{-1}H$ for which $a::b$.

Theorem 6.4 $a:\pi:b \;\Leftrightarrow\; Al_\pi\mathbb{B} = o.$

Corollary (i) $a :: b \iff Al_\pi\mathbb{B} = o$ for some π.[10]

Corollary (ii) $a : \pi : b \iff -b' : \pi^{-1} : -a'$.

For
$$(Al_\pi\mathbb{B})' = \mathbb{B}' l_\pi^{-1} A',$$
and A' is the $\mathbb{A}$ for $-a'$, and so forth.

Corollary (iii) $a :: b \iff -b' :: -a'$.

Theorem 6.5 If
$$[l \ \ a] = \begin{bmatrix} l_0 & 0 & \vdots & a_{00} & a_{01} \\ 0 & l_1 & \vdots & a_{10} & a_{11} \end{bmatrix},$$

$$[l \ \ a]_\pi = \begin{bmatrix} a_{00} & o & \vdots & l_0 & a_{01} \\ a_{10} & l_1 & \vdots & o & a_{11} \end{bmatrix} = [G \vdots H],$$

and

$$b = \begin{bmatrix} b_{00} & b_{01} \\ b_{10} & b_{11} \end{bmatrix},$$

then $a : \pi : b$ if and only if a_{00} is regular, and
$$b_{00} = a_{00}^{-1}, \qquad\qquad b_{01} = a_{00}^{-1} a_{01},$$
$$b_{10} = -a_{10} a_{00}^{-1}, \quad b_{11} = a_{11} - a_{10} a_{00}^{-1} a_{01}.$$

This is a consequence of Theorem 2, obtained by evaluating $b = G^{-1}H$.

The transformation from a to b is determined by the regular square submatrix a_{00} of a, so b is the *pivotal transform* of a, with a_{00} as *pivot*. The formula has been derived with consecutive partitions, matrix blocks being drawn from consecutive rows and columns, but it has sense independent of this.

The G, H in the theorem have a *canonical* arrangement, in regard to the regular square submatrix $\alpha = a_{00}$ of a in the leading position, but this arrangement also has an immateriality.

The permutation π swaps the columns in $[l \ a]$ containing a_{00} with the columns containing the unit submatrix l_0 of l which lies in the rows of α, l_1 being the complementary submatrix of l. The pivots α and permutations π are in a 1–1 correspondence by this

---

[10] Tucker states that M. M. Flood observed this form of condition for :: which permits a simple proof of Corollary (iii).

principle; and for every pivot there is a pivotal transformation of a producing a b such that $a : \pi : b$.

The following therefore is shown by the theorem.

Corollary (*i*) Let a_{rs} be a regular square submatrix of a, the sets r, s of rows and columns having complements i,j; let π be the permutation of the columns of $[1\ a]$ that swaps columns r of a in sequence with columns s of 1 keeping their order; and let b be the matrix given by

$$b_{rs} = a_{rs}^{-1}, \qquad b_{rj} = a_{rs}^{-1} a_{rj},$$
$$b_{is} = -a_{is} a_{rs}^{-1}, \quad b_{ij} = a_{ij} - a_{is} a_{rs}^{-1} a_{rj}.$$

Then

$$a : \pi : b.$$

For proceeding instead with consecutive partitions, and disregarding the order of the swap, given any regular square submatrix of a, and the permutation π of the columns of $[1\ a]$ which performs such a swap, and the G, H so obtained, there are unique permutations K of the rows and M, N of the columns of G, H that produce

$$G^* = K'GM, \quad H^* = K'HN$$

having the canonical arrangement. Then with $b = G^{-1}H$, as required for $a : \pi : b$, we have

$$b^* = G^{*-1}H^*$$
$$= (K'GM)^{-1}(K'HN) = M'(G^{-1}H)N$$
$$= M'bN.$$

The theorem shows an expression for b^* with reference to the canonical G^*, H^* and then $b = Mb^*N'$.

Corollary (*ii*) If $a : \pi : b$ and a_{00} is the pivotal submatrix of a corresponding to π, and a_{ij} are the submatrices of a in the partition so determined, then for some permutation matrices M, N of order m, n

$$M'bN = \begin{bmatrix} a_{00}^{-1} & a_{00}^{-1} a_{01} \\ -a_{10} a_{00}^{-1} & a_{11} - a_{10} a_{00}^{-1} a_{01} \end{bmatrix}.$$

Conversely, for any regular square submatrix a_{00} of a and permutation matrices M, N the b determined by this relation is such that $a :: b$.

From the formula for b such that $a :: b$, it appears that minor determinants of a, b have a correspondence, in which there is a proportionality, with ratio $\pm |\alpha|$, α being the pivot.

In Corollary (i), let i, j now denote subsets of the complements of r, s with the same number of elements. Then, for the minor determinant $|b_{ij}|$ of b on these rows and columns, by Schur's identity,

$$|a_{ri,sj}| = \begin{vmatrix} a_{rs} & a_{rj} \\ a_{is} & a_{ij} \end{vmatrix}$$

$$= |a_{rs}|\,|a_{ij} - a_{is}\,a_{rs}^{-1}\,a_{rj}| = |a_{rs}|\,|b_{ij}|,$$

and so

$$\frac{|b_{ij}|}{|a_{ri,sj}|} = \frac{1}{|a_{rs}|}.$$

If r, s have t elements, this being the *order* of the pivot α that produces b from a, and i, j have k elements, disjoint from these, a minor $|\alpha^*|$ of a of order $t + k$, above the pivot α, is here in correspondence with a minor $|\beta|$ of b of order k, the position of which is disjoint from that of the pivot; and

$$|\beta| / |\alpha^*| = 1/|\alpha|.$$

Let $\alpha \leftrightarrow \beta$ mean α, β are the matrices for minor determinants of a, b in this correspondence, and $a \underset{\alpha}{\rightarrow} b$ that b is produced from a by the pivot α.

Corollary (iii) $\alpha^* \leftrightarrow \beta .\Rightarrow. a \underset{\alpha^*}{\rightarrow} c \Leftrightarrow b \underset{\beta}{\rightarrow} c .$

It should be concluded that every pivotal transform of a matrix is identified by labels for its rows an columns, which, in any operation, undergo a swap in correspondence with the rows and columns of the pivot matrix. From this it follows that any operation with a pivot of order k can be accomplished by a series of k pivots of order 1, and *vice versa*.

Corollary (iv) If a is a regular square matrix, then

$$a :: a^{-1}, \text{ and } a \underset{a}{\rightarrow} a^{-1}.$$

Hence, the inversion of a regular square matrix of order n can be accomplished by a series of n pivots of order 1.

7
Determinants

7.1 Solution characteristics

Consider $x_0, x_1, \ldots, x_n \in K^n$. If $x_1, \ldots, x_n$ are independent, we have

$$x_0 = x_1 t_1 + \cdots + x_n t_n \qquad \text{(i)}$$

for unique $t_1, \ldots, t_n$. These are rational functions of the elements, as can be inferred from their determination by the elimination process, homogeneous linear in x_0. Transposition between x_0 and x_i replaces t_i by $1/t_i$ and t_j $(j \neq i)$ by $-t_j/t_i$. Hence there exists a polynomial δ, homogeneous linear and antisymmetric in n vector arguments, unique up to a constant multiplier, such that

$$t_i = \delta(x_1, \ldots, x_0, \ldots, x_n)/\delta(x_1, \ldots, x_i, \ldots, x_n).$$

The constant multiplier can be chosen to give it the value

$$\delta(1_1, \ldots, 1_n) = 1,$$

in respect to the elements of the fundamental base in K^n.

We now have the identity

$$x_0 \delta(x_1, \ldots, x_n) = x_1 \delta(x_0, \ldots, x_n) + \cdots + x_n \delta(x_1, \ldots, x_0) \quad \text{(ii)}$$

and will be able to conclude that the condition for $x_1, \ldots, x_n$ to be independent is that

$$\delta(x_1, \ldots, x_n) \neq 0.$$

Then the identity, subject to this condition, must also hold without it, by the principle of the irrelevance of algebraical inequalities, of Weyl (1939).

When the function δ is seen to have the Weierstrass characteristic properties identifying it with the *determinant*, from (ii) we have *Cramer's rule* for the solution of the simultaneous linear equations stated by (i).

7.2 Determinants

The function δ has the properties:

(I) linear form in each argument
$$\delta(\ldots,xt,\ldots) = \delta(\ldots,x,\ldots)t$$
$$\delta(\ldots,x+y,\ldots) = \delta(\ldots,x,\ldots) + \delta(\ldots,y,\ldots)$$

(II) antisymmetry
$$\delta(\ldots,x,\ldots,y,\ldots) = -\delta(\ldots,y,\ldots,x,\ldots),$$

(III) normalization on a base
$$\delta(I_1,\ldots,I_n) = 1,$$

We have inferred the existence of a function with these properties. Regarding a matrix as formed by the set of vectors provided by its columns, these are the three *Weierstrass Characteristic Properties* for the determinant of a matrix, by which it is completely defined. The determinant being so identified, from the last section we already have *Cramer's Rule* for the solution of linear equations[11]. The rule may be recovered from the Weierstrass properties. Also, from these properties, while deriving the formula for the determinant as a polynomial, at the same time there will be a proof of the *Multiplication Theorem* for determinants, that the determinant of a product of matrices is the product of the determinants.

As a consequence of (II),
$$\delta(\ldots,x,\ldots,x,\ldots) = 0, \qquad\qquad\text{(i)}$$
that is, $\delta = 0$ if two arguments are equal. Now with (I),
$$\delta(\ldots,x+y,\ldots,y,\ldots)$$
$$= \delta(\ldots,x,\ldots,y,\ldots) + \delta(\ldots,y,\ldots,y,\ldots)$$
$$= \delta(\ldots,x,\ldots,y,\ldots).$$

Hence, δ is unchanged when one argument is replaced by its sum with another. We now have, from (I) and (II), that the effect of

---

[11] Tucker mentioned in a letter (December 30, 1987) about G. R. Kirchhoff (1847) "reaching a Cramer solution without determinants". He provided me with the translation of Kirchoff by J. G. Kemeny (1946), and his own note "On Kirchoff's laws …" (1948).

elementary operations on its arguments is to multiply δ by a non-zero constant. Again by (II), if π is any permutation of $1, \ldots, n$,

$$\delta(x_{\pi 1}, \ldots, x_{\pi n}) = \chi_\pi \delta(x_1, \ldots, x_n), \tag{ii}$$

where χ_π is the alternating character of π, 1 or -1 according as π is an odd or even permutation.

The determinant of a matrix b is the polynomial in its elements given by

$$|b| = \sum_\pi \chi_\pi b_{\pi 1,1} \ldots b_{\pi n,n}, \tag{iii}$$

where the summation is over all permutations π of $1, \ldots, n$. Considered as a function of its n columns, it is obvious that the determinant has the Weierstrasse properties, shared by our function δ.

When it is shown, as a consequence of (I) and (II) for δ, that

$$\delta(ab) = \delta(a)|b|,$$

it will follow, by taking $a = 1$, that

$$\delta(b) = \delta(1)|b|,$$

and hence, with $\delta(1) = 1$ from (III), that

$$\delta(b) = |b|.$$

Then, by substitution for δ above, it will have been shown at the same time that

$$|ab| = |a||b|,$$

which is the *multiplication theorem* for determinants. Since δ enters as any function with the Weierstrass properties (I), (II) and (III), shared by the determinant, it will also have been shown that *the determinant is the unique function with the Weierstrass properties*.

Theorem 2.1 (I) and (II) imply

$$\delta(ab) = \delta(a)|b|.$$

With the columns of a denoted $a_j = a_{j)}$, the columns of ab are

$$(ab)_{j)} = ab_{j)} = a_1 b_{1j} + \ldots + a_n b_{nj}.$$

Hence $\delta(ab)$ has the expression

$$\delta(a_1 b_{11} + \cdots + a_n b_{n1}, \ldots, a_1 b_{1n} + \cdots + a_n b_{nn}).$$

Expanding by repeated application of (I) and (II), this becomes

$$\sum_i \delta(a_{i_1}, \ldots, a_{i_n}) b_{i_1,1} \ldots b_{i_n,n},$$

where the indices in the set $i = (i_1, \dots , i_n)$ range independently over $1, \dots , n$. But by (i), the terms where i is not a permutation of $1, \dots , n$ vanish. Then by (ii) and (iii), we are left with $\delta(a)|b|$. QED

Now we proceed from the characteristic properties of a determinant, and conclusions that have been drawn directly from them.

Theorem 2.2 The determinant of a matrix is a prime polynomial in its elements.

With a factorization $|x| = f(x)g(x)$, if x_{ij} is an element that occurs in f then, from the formula for the determinant, no elements of the form x_{ik} or x_{kj}, for some k, can occur in g. If now x_{rs} is an element that occurs in g, so $r \neq i$ and $s \neq j$, then no elements of the form x_{rk} or x_{ks} can occur in f. In that case x_{is} and x_{rj} are elements that cannot occur in either f or g. But all elements occur in the determinant, so there is a contradiction. Hence the factorization is impossible. QED

This theorem, and the proof, are in Aitken (1942, p. 37).

Theorem 2.3 If $x_1, \dots , x_n$ are dependent, then

$$|x_1 \dots \ x_n| = 0.$$

For then one is a combination of the others, say

$$x_1 = x_2 t_2 + \cdots + x_n t_n,$$

and then, substituting for x_1 and expanding by linearity,

$$|x_1 \dots \ x_n| = |x_2 x_2 \dots \ x_n|t_2 + \cdots + |x_n x_2 \dots \ x_n|t_n = 0,$$

all terms vanishing as a consequence of (i).

Theorem 2.4 For any $n + 1$ elements in K^n, there is the identity

$$x_0|x_1 \dots \ x_n| = x_1|x_0 \dots \ x_n| + \cdots + x_n|x_1 \dots \ x_0|.$$

At first suppose

$$|x_1 \dots \ x_n| \neq 0. \tag{i}$$

Then, by Theorem 2.3, $x_1, \ \dots , x_n$ are independent while, by the dimension theorem, $x_0, x_1, \dots , x_n$ are dependent. Hence x_0 is a unique combination of $x_1, \dots , x_n$,

$$x_0 = x_1 t_1 + \cdots + x_n t_n.$$

Then

$$|x_0 x_2 \ldots x_n|$$

$$= |x_1 x_2 \ldots x_n| t_1 + |x_2 x_2 \ldots x_n| t_2 + \cdots + |x_n x_2 \ldots x_n| t_n$$

$$= |x_1 x_2 \ldots x_n| t_1,$$

so that

$$t_1 = |x_0 \ldots x_n| / |x_1 \ldots x_n|,$$

and similarly

$$\ldots, t_n = |x_1 \ldots x_0| / |x_1 \ldots x_n|,$$

which proves the identity subject to the inequality (i). But then it must be valid also without this restriction, by Weyl's irrelevance principle.

Theorem 2.5 If $x_1, \ldots, x_n$ are independent, then

$$|x_1 \ldots x_n| \neq 0.$$

For then they form a base in K^n, and any n vectors have expressions as combinations of them, given by the columns of $y = xa$, for some matrix a. By the multiplication theorem, if $|x| = 0$, then

$$|y| = |x||a| = 0,$$

showing that the determinant is identically zero, which is impossible, since anyway $|I| = 1$. Hence $|x| \neq 0$.

Theorem 2.6 A necessary and sufficient condition for any elements $x_1, \ldots, x_n \in K^n$ to be independent, and so form a base in K^n, is that

$$|x_1 \ldots x_n| \neq 0,$$

and then the unique coordinates $t_1, \ldots, t_n$ of any element $x_0 \in K^n$ relative to this base, such that

$$x_0 = x_1 t_1 + \cdots + x_n t_n,$$

are given by

$$t_1 = |x_0 \ldots x_n| / |x_1 \ldots x_n|,$$

$$\cdots \quad \cdots \quad \cdots$$

$$t_n = |x_1 \ldots x_0| / |x_1 \ldots x_n|.$$

This comes from Theorems 2.3-2.5, and is *Cramer's Rule*.

7.3 Adjoint and inverse

For any matrix a, and the fundamental base vectors $1_j = 1_{j)}$, given by the columns of the unit matrix 1, we may form the matrix A, the *adjoint* of a, which is also denoted adj $a = A$, with elements

$$A_{ij} = |a_1 \ldots 1_j \ldots a_n|,$$

obtained by making 1_j replace a_i in $|a|$. By Theorem 2.6,

$$1_j|a| = \sum_i a_i A_{ij},$$

that is,

$$1|a| = aA.$$

Then, provided $|a| \neq 0$, the matrix $b = A|a|^{-1}$ is defined, and such that $ab = 1$, so it is a right inverse of a. Therefore it is also a left inverse, and hence the unique inverse of a. It follows that a is regular, with inverse given by

$$a^{-1} = A|a|^{-1},$$

such that

$$aa^{-1} = 1 = a^{-1}a.$$

Conversely, if a is a regular matrix, having an inverse, then the multiplication theorem for determinants shows that $|a| \neq 0$, and moreover that

$$|a^{-1}| = |a|^{-1}.$$

With any given matrix a, and its adjoint A, we have

$$|a| = \sum_j a_{ij} A_{ji}$$

and

$$0 = \sum_j a_{ij} A_{jk} \ (k \neq i).$$

The first case shows the *Laplace expansion* of the determinant by the elements a_{ij} in row i, A_{ji} being the cofactor of a_{ij} in a. In the other case, we have the Laplace expansion on row i by *alien factors*, these being the cofactors associated with another row $k \neq i$. The result is 0, since this is the Laplace expansion of the determinant, which has two rows identical and so must vanish, obtained from $|a|$ by replacing row i by row k.

For computation of the inverse of a square matrix a of order n, one method is to form the table

$$a \vdots 1$$

where the unit matrix I is adjoined, and perform a maximal replacement with pivots restricted to the first n columns. If the result is

$$I : b$$

then b is the inverse of a. Otherwise a is established as of rank $r < n$, and so without an inverse.

7.4 Derived system

For matrices a, b both of which are of order $n \times n$, the multiplication theorem for determinants asserts that

$$|ab| = |a||b|.$$

An extension of this multiplication theorem provides the basis for the Binet-Cauchy theorem on derived matrices. First consider a, b of order $m \times n$, $n \times m$ where $m \leq n$, and the $m \times m$ determinant ab. For any set of m distinct indices

$$k = (k_1, \ldots, k_m), \quad 1 \leq k_1 < \cdots < k_m \leq n,$$

taken from $1, \ldots, n$, $a_{k)}$ will denote the submatrix of a on those columns, and $b_{(k}$ the submatrix of b on those rows. There are $\binom{n}{m}$ possibilities for k, and so that number of such submatrices. They are all of order $m \times m$, each with a determinant. The determinant of ab is found to be the sum of products of corresponding mth order subdeterminants from a and b.

If $m = n$, in which case a, b have ordinary determinants, and $k = (1, \ldots, n)$ is the only possibility for k, there is just one such product, namely ab, so we have a generalization of the ordinary multiplication theorem.

Theorem 4.1 $\qquad |ab| = \sum_k |a_{(k}||b_{k)}|.$

Since

$$(ab)_{ij} = \sum_h a_{ih} b_{hk},$$

$|ab|$ has the expression

$$|ab| = \sum_\pi \chi_\pi \left(\sum_{h_1} a_{1,h_1} b_{h_1, \pi 1} \right) \cdots \left(\sum_{h_m} a_{1,h_m} b_{h_m, \pi m} \right).$$

In the rearrangement, terms where the h_i are not all distinct would vanish. Then with h_i distinct, a permutation ρ will put them in ascending order, as $k_i = \rho h_i$. Correspondingly, the πi become rearranged to σi, where $\sigma = \rho^{-1}\pi$. Hence the expression becomes

$$\textstyle\sum_k \sum_{\rho,\sigma} \chi_{\rho\sigma} \left(a_{1,\rho k_1} \cdots a_{m,\rho k_m} \right) \left(b_{k_1,\sigma 1} \cdots b_{k_m,\sigma m} \right),$$

and, since $\chi_{\rho\sigma} = \chi_\rho \chi_\sigma$, this becomes

$$\textstyle\sum_k \left(\sum_\rho \chi_\rho a_{1,\rho k_1} \cdots a_{m,\rho k_m} \right) \left(\sum_\sigma \chi_\sigma b_{k_1,\sigma 1} \cdots b_{k_m,\sigma m} \right),$$

as required.

Suppose now we have a, b of order $p \times n$, $n \times q$ and so with a product ab, of order $p \times q$. For $m \le p, n, q$ let $a_{(i}$ denote the $m \times n$ submatrix of a formed from the m rows with indices in the set

$$i = (i_1, \dots, i_m), \quad 1 \le i_1 < \cdots < i_m \le p,$$

and similarly with $b_{j)}$ drawn from columns of b. Also, a_{ik} can denoted the submatrix of a on sets i, k of rows and columns, and similarly with b_{kj} for b.

According to the theorem,

$$|a_{(i} b_{j)}| = \textstyle\sum_k |a_{ik}| \, |b_{kj}|.$$

In Cauchy's notation for *derived matrices* which will be described, this shows that

$$(ab)^{(m)} = a^{(m)} b^{(m)}, \tag{i}$$

which states the *Binet-Cauchy Theorem*, that the mth derived matrix of a product is the product of the derived matrices.

With any matrix a, or order $p \times q$, and $m \le p, q$, the elements of the mth derived matrix $a^{(m)}$ are the mth order determinants taken from rows and columns of a. In present notation, these are a_{ij}, where i, j are sets of indices for m rows and columns. With Cauchy, index sets, and hence $m \times m$-submatrices, are ordered in a series by the order of last differing elements (see Appendix, §2). The series position of any index set i for rows of a is given by

$$r = 1 + \textstyle\sum_s \binom{i_s - 1}{s},$$

and similarly a position s is determined for j; and the sets i, j can be recovered from the indices r, s. Then

$$a_{rs}^{(m)} = |a_{ij}|$$

is the r, sth element of the mth *derived matrix* $a^{(m)}$ of a. The present theorem tells that

$$(ab)_{rs}^{(m)} = a_{(r}^{(m)} b_{s)}^{(m)},$$

which is (i).

For a square matrix a, with inverse a^{-1},

$$(a^{-1})^{(m)} = (a^{(m)})^{-1}.$$

Also, if $a^{[m]}$ is the transpose of the matrix obtained by replacing the elements of $a^{(m)}$ by their cofactors in $|a|$, then

$$a^{(m)}a^{[m]} = |a|I.$$

This defines the *m*th *adjoint derived matrix* $a^{[m]}$. Then adj $a = a^{[1]}$ for the usual adjoint, and moreover,

$$(\text{adj } a)^{(m)} = a^{[m]}|a|^{m-1},$$

which expresses "the celebrated theorem of Jacobi (1834) concerning the minors of the adjugate"(Aitken (1942), p. 98.).

7.5 Double determinants

In §8.1 there is the identity

$$\left| -\begin{bmatrix} o & l & p \\ l & b & h \\ q & k & c \end{bmatrix} \right| = |-(-c + kp + qh - qbp)|,$$

where the matrices b, o are square of the same order. The negative signs serve to avoid having to make explicit reference to matrix orders, and display the form of a twice repeated determinant operation. When the matrices involved are all of order unity, this identity becomes vacuous, as a statement of definition of a determinant. Otherwise, there does appear to be a determinant operation involving matrices as elements, though to make this interpretation it is necessary that the unit matrices be freed of a fixed order, to make sense of the multiplications they would enter.

But this determinant can be taken without such qualifications when the matrices are all of the same order, some convention having been decided for the order of multiplication of factors in each terms, should those factors not be commutative. The identity then shows a twice-repeated determinant operation on the right as equivalent to a simple one on the left. This suggests the theorem to be dealt with now, which has a number of applications (as mentioned in the introduction).

A case of the theorem comes immediately from Schur's identity

$$\begin{vmatrix} a & u \\ v & b \end{vmatrix} = |a||b - va^{-1}u|$$

For, from this, if $av = va$, then

$$\begin{vmatrix} a & u \\ v & b \end{vmatrix} = |a|\,|b - va^{-1}u| = |ab - vu| = \left\|\begin{matrix} a & u \\ v & b \end{matrix}\right\|.$$

Lemma If $|a \vdots b|$ is a determinant partitioned in its columns, and t is a square matrix of order equal to the column order of a, then

$$|at \vdots b| = |a \vdots b|\,|t|.$$

When a is replaced by at, minors on the columns of a are multiplied by the determinant of t. But, by Laplace's theorem, a determinant is a linear function of the minors on any set of columns, and hence the required conclusion.

A *composite matrix*, of *order n* and *degree m*, is an nth order square matrix whose elements, its *components*, are square matrices of order m. It has a *ground* matrix, the simple matrix of order mn from which it can be derived by partitioning. With a *commutative composite* matrix, the components all commute. Then one may take its determinant, the result being a matrix, and the determinant of this is the *double-determinant* of the composite matrix, while the *ground determinant* is the determinant of the ground matrix.

Theorem 5.1 The double determinant of a commutative composite matrix is equal to its ground determinant.

The proof is by induction on the order n of the composite matrix. The case $n = 1$ being verified, take as inductive hypothesis the case for $n - 1$, and consider a composite matrix A of order n, with component matrices A_{ij} which commute, and ground matrix a. It is required to prove that

$$\|A\| = |a|.$$

By definition,

$$|A| = A_{11}|B_1| - A_{12}|B_2| + \cdots + (-1)^{n-1}A_{1n}|B_n|,$$

this being the expansion of the determinant of A on the components in the first row, where B_i denotes the composite matrix of order $n - 1$ formed from all the rows of A except the first, and all columns except the ith. If b_i is the ground matrix of B_i, then, by the inductive hypothesis,

$$\|B_i\| = |b_i|.$$

By the Lemma,

$$|a|\,\|B_1\| = \begin{vmatrix} A_{11}|B_1| & A_{12} & \cdots & A_{1n} \\ \cdots & \cdots & \cdots & \cdots \\ A_{n1}|B_1| & A_{n2} & \cdots & A_{nn} \end{vmatrix}$$

Then by elementary operations on the columns, which leave the value of the determinant unchanged, this expression becomes

$$\begin{vmatrix} A_{11}B_1 - \cdots + (-1)^{n-1}A_{1n}B_n & A_{12} & \cdots & A_{1n} \\ A_{n1}B_1 - \cdots + (-1)^{n-1}A_{nn}B_n & A_{n2} & \cdots & A_{nn} \end{vmatrix}$$

All components in the first column are null, exception the first, which is $|A|$. For they are all expansions of composite determinants which have two of their rows identical, and so vanish as a consequence of the commutativity assumptions, except the first which is an expansion of A. Hence the expression for the determinant reduces to

$$\begin{vmatrix} |A| & \vdots & A_{12} & \cdots & A_{1n} \\ O & \vdots & & b_1 \end{vmatrix} = |b_1||A|.$$

But, by hypothesis, $|b_1| = ||B_1||$, so now

$$|b_1|\,||A|| = |b_1||a|.$$

Hence, provided $|b_1| \neq 0$, we have the required conclusion

$$||A|| = |a|.$$

But this inequality is irrelevant to this conclusion, by the Principle of the Irrelevance of Algebraic Inequalities, of Weyl (1939), so the proof by induction is complete.

7.6 Alternants

The function now to be considered has an association with the affine space, analogous to the way the determinant is associated with the linear space. The *alternant* is defined as the unique function σ of $n + 1$ vector arguments $x_0, x_1, \ldots, x_n \in K^n$ with the properties

(I*) linearity in each argument

$$\sigma(\ldots, xs + yt, \ldots) = \sigma(\ldots, x, \ldots)s + \sigma(\ldots, y, \ldots)t,$$

for $s + t = 1$,

(II*) invariance under translation

$$\sigma(x + d, y + d, \ldots) = \sigma(x, y, \ldots),$$

(III*) antisymmetry

$$\sigma(\ldots, x, \ldots, y, \ldots) = -\sigma(\ldots, y, \ldots, x, \ldots),$$

(IV*) normalization on a regular simplex

$$\sigma(o, l_1, \ldots, l_n) = 1.$$

For the unique characterization of σ by these properties, we have that if

$$\delta(x_1, \ldots, x_n) = \sigma(o, x_1, \ldots, x_n), \qquad (i)$$

then

$$\sigma(x_0, x_1, \ldots, x_n) = \delta(x_1 - x_0, \ldots, x_n - x_0), \qquad (ii)$$

where the function δ, determined from σ and from which σ is recovered, has the Weierstrass properties, (I)–(III) in section 7.2, that identify it with the determinant.

From (II*), by taking $d = -x_0$ we have (ii), for δ given by (i). It remains to see that, with σ having the considered properties, δ given by (i) has the properties of the determinant.

By (i), also δ has the linearity property (I*). But from the antisymmetry (III*) for σ, we have, beside from (i) that also δ has this property, the further conclusion that $\sigma = 0$ if $x_0 = x_i$, and hence, with (ii), that $\delta = 0$ if any of its arguments is null. This shows that δ, already linear, is homogeneous linear in each argument, as required by (I) and (II). Finally, from (IV*), $\delta(l) = 1$, which is (III). QED

Theorem 6.1 The properties of the alternant σ imply, and are implied by, the identity

$$\sigma(x_0, x_1, \ldots, x_n)$$
$$= \delta(x_1, \ldots, x_n) - \delta(x_0, \ldots, x_n) - \cdots - \delta(x_1, \ldots, x_0),$$

where δ is the determinant.

With the above, this formula comes from expanding the right side of (ii), using the properties of the determinant. Also, from the properties of the determinant the function σ, given in this way in terms of the determinant, has all the properties of the alternant. QED

Corollary For any matrix $a \in K_n^n$,

$$\sigma(ax_0, \ldots, ax_n) = \delta(a)\sigma(x_0, \ldots, x_n).$$

This follows from the multiplication theorem for determinants. It combines with the property (II*), of invariance under translations, to show the characteristic of σ under affine transformations.

Theorem 6.2 For any $n+1$ elements $x_0, \ldots, x_n \in K^n$, and further element $y \in K^n$, there are the identities

$$y\sigma(x_0, \ldots, x_n) = x_0\sigma(y, \ldots, x_n) + \cdots + x_n\sigma(x_0, \ldots, y),$$

and

$$\sigma(x_0, \ldots, x_n) = \sigma(y, \ldots, x_n) + \cdots + \sigma(x_0, \ldots, y).$$

The second identity is obtained from the formula of Theorem 6.1 by a summation in respect to every set of $n+1$ elements taken from the $n+2$. Thus, taking the formulae for

$$\sigma(y, \ldots, x_n), \ldots, \sigma(x_0, \ldots, y)$$

in terms of the determinant and adding them, terms cancel and we are left with the expression for $\sigma(x_0, \ldots, x_n)$ in terms of the determinant.

By Theorem 7.2.4, there is the identity

$$y\delta(x_1, \ldots, x_n) = x_1\delta(y, \ldots, x_n) + \cdots + x_n\delta(x_1, \ldots, y),$$

and then other identities where x_0 replaces each of $x_1, \ldots, x_n$. By subtracting the sum of these further identities from the first, the first identity of the present theorem is obtained. QED

Cramer's well known rule for solving simultaneous linear equations amounts to a formula for finding the coordinates of any element $b \in K^n$ relative to a base $a_1, \ldots, a_n$ in K^n. For in asking to solve $ax = b$, for given regular matrix $a \in K_n^n$, we are seeking an expression

$$b = a_1x_1 + \cdots + a_nx_n$$

for b as a linear combination of the independent columns $a_1, \ldots, a_n$ of a.

The affine coordinates of an element with a given simplex of reference, dealt with in §4.4.6, have a formula that is analogous to Cramer's rule, understood as a formula for linear coordinates. It involves the alternant in just the same way as Cramer's rule involves the determinant.

Theorem 6.2 is going to provide this rule. It only needs to be recognized that

$$\sigma(a_0, a_1, \ldots, a_n) \neq 0$$

is the condition for the affine independence of any $n+1$ elements, or for these to be the vertices of a regular simplex, just as

$$\delta(a_1, \ldots, a_n) \neq 0$$

provides the condition for the linear independence of any n elements, or for these to be a base.

But, with reference to §4.5, and in view of the corresponding role for δ, we see this for σ immediately from (ii) of the last section.

Theorem 6.3 A necessary and sufficient condition for any $n + 1$ elements $a_0, \ldots, a_n \in K^n$ to be the vertices of a regular n-simplex in K^n is that

$$\sigma(a_0, \ldots, a_n) \neq 0.$$

Then the unique affine coordinates $t_0, \ldots, t_n$ of any element $x \in K^n$ with this as simplex of reference, for which

$$a_0 t_0 + \cdots + a_n t_n = x,$$

$$t_0 + \cdots + t_n = 1,$$

are given by

$$t_0 = \sigma(x, \ldots, a_n)/\sigma(a_0, \ldots, a_n),$$

$$\cdots \quad \cdots \quad \cdots \quad \cdots$$

$$t_n = \sigma(a_0, \ldots, x)/\sigma(a_0, \ldots, a_n).$$

Any set of p elements $x_1, \ldots, x_p$ is given as an ordered set S by an expression

$$S = (x_1, \ldots, x_p).$$

It is given as an oriented set when the order is determinate to within an even permutation. Hence there are two opposite orientations, and the set is oriented when these are distinguished as positive and negative. If S is oriented, $-S$ would be the set with opposite orientation, so if π is any permutation of $1, \ldots, p$ then

$$(x_{\pi 1}, \ldots, x_{\pi p}) = \chi_\pi(x_1, \ldots, x_p),$$

where χ_π is the alternating character of π, 1 or -1 according as π is an odd or even permutation.

Any base in $\mathfrak{R}^n$ can be taken to be oriented, positive or negative, by the order in which its elements are given, in particular the fundamental base. By the antisymmetry of the determinant and its non-vanishing on any base, it can be taken that a base has positive orientation when in an order for which the determinant is positive. The determinant has been normalized so as to make the fundamental base positively oriented. The distinction between bases that have the same or opposite orientation is that those with the same can be connected by a continuous path of bases, whereas those with opposite cannot. That those with opposite orientation cannot be so connected is quite immediate, though the converse is not.

Similarly by the antisymmetry of σ, any regular n-simplex in $\mathfrak{R}^n$ may be oriented by orienting its $n + 1$ vertices, positive or negative according to the sign of $\sigma(a_0, a_1, \dots, a_n)$. The normalization of σ is such as to give positive orientation to the simplex $o, l_1, \dots, l_n$ obtained by taking the origin with the fundamental base in $\mathfrak{R}^n$. As with bases, the regular n-simplices with like orientation are connected by a continuous path of regular simplices, and those with opposite are not.

8
Determinants and Matrices

8.1 Partitioned determinants

Theorem 1.1

$$\begin{bmatrix} 1 & o \\ -va^{-1} & 1 \end{bmatrix} \begin{bmatrix} a & u \\ v & b \end{bmatrix} \begin{bmatrix} 1 & -a^{-1}u \\ o & 1 \end{bmatrix} = \begin{bmatrix} a & o \\ o & b-va^{-1}u \end{bmatrix},$$

provided a is regular.

The identity is seen directly by evaluating the product on the left.

Corollary (i)
$$\begin{vmatrix} a & u \\ v & b \end{vmatrix} = |a||b - va^{-1}u|.$$

provided a is regular.

This is Schur's identity, shown already in Chapter 6, §5, by another method. It is obtained here by taking determinants in the identity, and using the determinant multiplication theorem, with the observation that a triangular matrix with units on the diagonal has determinant 1.

A simpler statement for the case where b is 1×1, or a single element of the matrix, in which case u, v are vectors, is just that

$$\begin{vmatrix} a & u \\ v & b \end{vmatrix} = |a|(b - va^{-1}u).$$

Corollary (ii)
$$\begin{vmatrix} a & u \\ v & o \end{vmatrix} = (-1)^m |va^{-1}u|,$$

provided a is regular, where a is $n \times n$, and u, v are $n \times m$, $m \times n$.

Corollary (iii) $\begin{vmatrix} a & u \\ v & b \end{vmatrix} = |a|b - vAu,$

where u, v are vectors, and $A = \text{adj } a$.

First we have it when a is regular, since $a^{-1} = |a|^{-1}A$, and then the qualification can be dropped, by the Algebraic Irrelevance Principle of Weyl (1939).

In § 5.3, the rank of a matrix has been defined from the coincidence of row rank and column rank, as their common value. If a is of rank r, then any regular $r \times r$ submatrix p is a critical submatrix, in respect to which a has a critical decomposition.

A *minor* of a matrix is the determinant of a square submatrix. It is a *critical minor* if it is non-zero, but every minor of higher order containing it is zero (there is no confusion when the term refers also to the square submatrix).

It will be seen that the order of any critical minor of a matrix coincides with the rank. Clearly these are just the critical submatrices of § 5.3. We now give a new proof, using determinants, of the critical decomposition formula for a matrix, Theorem 3.4 there; it confirms the correspondence of critical minors with critical submatrices.

Corollary (iv)   If U, V are the submatrices of a matrix a of the rows and columns through a critical minor p, then

$$a = Vp^{-1}U.$$

With

$$a = \begin{bmatrix} p & u \\ v & d \end{bmatrix}$$

where p is $r \times r$ and regular, the minors of order $r + 1$ are

$$\begin{vmatrix} p & u_j \\ v_i & d_{ij} \end{vmatrix} = |p|(d_{ij} - v_i p^{-1} u_j)$$

by Corollary (ii), where v_i, u_j are row i and column j of u, v. Since p is critical, these are all zero, so that

$$d_{ij} = v_i p^{-1} u_j,$$

that is,

$$d = vp^{-1}u,$$

equivalently,

$$a = \begin{bmatrix} p \\ v \end{bmatrix} p^{-1} [p \ u].$$

Hence, taking

$$U = [p \ u], \quad V = \begin{bmatrix} p \\ v \end{bmatrix},$$

we have the required conclusion.

It comes directly from the formula so obtained that the order of the critical minor coincides with the row rank and the column rank. Hence all critical minors have the same order, given by the rank.

Consider an $n \times n$ matrix a, leadingly bordered by $m < n$ independent rows and columns, to form the *bordered matrix*

$$A = \begin{bmatrix} o & u \\ v & a \end{bmatrix},$$

of order $(n + m) \times (n + m)$, the u, v being of order $m \times n$, $n \times m$ and rank m. With partitions

$$u = [p \ r], \quad v = \begin{bmatrix} q \\ s \end{bmatrix},$$

it may be assumed that p, q are square and regular. Correspondingly,

$$a = \begin{bmatrix} b & h \\ k & d \end{bmatrix}$$

and

$$A = \begin{bmatrix} o & p & r \\ q & b & h \\ s & k & d \end{bmatrix}.$$

Now with

$$e = p^{-1}r, \quad f = sq^{-1},$$

consider also

$$G = \begin{bmatrix} o & l & e \\ l & b & h \\ f & k & d \end{bmatrix},$$

and

$$H = [-f \ l] \begin{bmatrix} b & h \\ k & d \end{bmatrix} \begin{bmatrix} -e \\ l \end{bmatrix} = fbe - fh - ke + d.$$

Lemma $\begin{bmatrix} o & l \\ l & b \end{bmatrix}^{-1} = \begin{bmatrix} -b & l \\ l & o \end{bmatrix}, \quad \begin{vmatrix} o & l \\ l & b \end{vmatrix} = (-1)^m.$

The inverse is verified directly. For the determinant, by m^2 transpositions of columns, it becomes

$$(-1)^{m^2}\begin{vmatrix} I & o \\ b & I \end{vmatrix} = (-1)^{m^2}.$$

But $(-1)^{m^2} = (-1)^m$.

Theorem 1.2 $\qquad |G| = (-1)^m |H|.$

By Schur's identity together with the Lemma,

$$|G| = \begin{vmatrix} o & I \\ I & b \end{vmatrix} \left| d - [f \ k] \begin{bmatrix} o & I \\ I & b \end{bmatrix}^{-1} \begin{bmatrix} e \\ h \end{bmatrix} \right|$$

$$= (-1)^m \left| d - [f \ k] \begin{bmatrix} -b & I \\ I & o \end{bmatrix} \begin{bmatrix} e \\ h \end{bmatrix} \right|$$

$$= (-1)^m |d + fbe - fh - ke| = (-1)^m |H|.$$

Corollary $\qquad |A| = (-1)^m |p| \, |q| \, |H|.$

For

$$|p| \, |q| \begin{vmatrix} o & I & e \\ I & b & h \\ f & k & d \end{vmatrix} = \begin{vmatrix} o & p & r \\ q & b & h \\ s & k & d \end{vmatrix} = \begin{vmatrix} o & u \\ v & a \end{vmatrix},$$

that is,

$$|p| \, |q| \, |G| = |A|.$$

8.2 Partitioned inverse

Theorem 2.1 If a is regular, then

$$\begin{bmatrix} a & u \\ v & b \end{bmatrix}$$

is regular if and only if

$$b - va^{-1}u$$

is regular, and then

$$\begin{bmatrix} a & u \\ v & b \end{bmatrix}^{-1} = \begin{bmatrix} A & U \\ V & B \end{bmatrix},$$

where

$$B = (b - va^{-1}u)^{-1},$$

$$A = a^{-1} + a^{-1}uBva^{-1},$$

$$U = -a^{-1}uB, \quad V = -Bva^{-1}.$$

The first part comes from Schur's identity, Corollary (i) in §1. The remainder follows from Theorem 1.1, which shows that

$$\begin{bmatrix} a & u \\ v & b \end{bmatrix}^{-1} = \begin{bmatrix} 1 & -a^{-1}u \\ 0 & 1 \end{bmatrix} \begin{bmatrix} a^{-1} & o \\ o & B \end{bmatrix} \begin{bmatrix} 1 & o \\ -va^{-1} & 1 \end{bmatrix}.$$

The square submatrices a and B in complementary positions in the matrix and its inverse have a reciprocal relation, making them *inverse complements, a* being associated with *B* in just the same way that *B* is associated with *a.*

The relation between the blocks in the matrix and its inverse is stated more symmetrically by

$$B^{-1} = b - va^{-1}u, \quad a^{-1} = A - UB^{-1}V$$

$$UB^{-1} + a^{-1}u = o, \quad B^{-1}V + va^{-1} = o.$$

The defining conditions for these relations are

$$aA + uV = 1, \quad aU + uB = o,$$

$$vA + bV = o, \quad vU + bB = 1,$$

or the same obtained with exchange between the matrix and its inverse, from their commutativity.

Corollary (i)   If *a* is regular, then

$$\begin{bmatrix} a & u \\ v & o \end{bmatrix}$$

is regular if and only if $va^{-1}u$ is regular, and then

$$\begin{bmatrix} a & u \\ v & o \end{bmatrix}^{-1} = \begin{bmatrix} A & U \\ V & B \end{bmatrix},$$

where

$$A = a^{-1}(a - u(va^{-1}u)^{-1}v)a^{-1},$$

$$B = -(va^{-1}u)^{-1},$$

$$U = a^{-1}u(va^{-1}u)^{-1}, \quad V = (va^{-1}u)^{-1}va^{-1}.$$

Consequently,

$$vU = 1, \quad Vu = 1.$$

from which it follows that Uv, uV are idempotents, that is,

$$(Uv)^2 = Uv, \quad (uV)^2 = uV.$$

Hence they are projectors, touched in § 2.2, so

$$Aa = 1 - Uv, \quad aA = 1 - uV$$

are the complementary projectors. Moreover,

$$uV = u(va^{-1}u)^{-1}va^{-1}, \quad Uv = a^{-1}u(va^{-1}u)^{-1}v.$$

If A is symmetric, so that $v = u'$ and $V = U'$, then also

$$uV = (Uv)'.$$

A further special case of $b = o$, where also u, v are square and regular, has simplifications. In this case we have

$$A = o, \quad B = -u^{-1}av^{-1},$$
$$U = v^{-1}, \quad V = u^{-1},$$

without requirement of the regularity of a.

Corollary (ii) If u, v are square and regular, then

$$\begin{bmatrix} a & u \\ v & o \end{bmatrix}^{-1} = \begin{bmatrix} o & v^{-1} \\ u^{-1} & -u^{-1}av^{-1} \end{bmatrix}.$$

Another special case of interest is the following.

Corollary (iii) $\quad \begin{bmatrix} a & u \\ o & b \end{bmatrix}^{-1} = \begin{bmatrix} a^{-1} & -a^{-1}ub^{-1} \\ o & b^{-1} \end{bmatrix}.$

An *upper triangular* matrix u is such that $u_{ij} = 0$ for $i > j$, and it is a *unit upper triangular* matrix if also $u_{ii} = 1$ for all i; and similarly for a *lower triangular* matrix.

Theorem 2.2 If u is unit upper triangular, then so is u^{-1}.

The proof is by induction on the order of u. With

$$U = \begin{bmatrix} u & x \\ o & 1 \end{bmatrix}$$

as a unit upper triangular matrix of order $n + 1$, u is unit upper triangular of order n, and so, by hypothesis, u^{-1} is unit upper triangular. But, by Corollary (iii),

$$U^{-1} = \begin{bmatrix} u^{-1} & y \\ o & 1 \end{bmatrix},$$

and hence also this is unit upper triangular. QED

8.3 Principal minors

The *leading principal minor* of order m of a matrix a is

$$\delta_m = \begin{vmatrix} a_{11} & \cdots & a_{1m} \\ \vdots & \ddots & \vdots \\ a_{m1} & \cdots & a_{mm} \end{vmatrix}.$$

Theorem 3.1 For a square matrix a of order n, with δ_m as its leading principal minor of order m, if

$$\delta_m \neq 0 \ (m = 1, \dots, n)$$

then there exist unit upper and lower triangular matrices u, v such that

$$vau = d,$$

where d is the the diagonal matrix with elements given by

$$d_m = \delta_m/\delta_{m-1} \ (m = 1, \dots, n),$$

with $\delta_0 = 1$.

The proof is by induction on the order of the matrix. The case $n = 1$ being verified, the inductive hypothesis is for $n \geq 1$. Consider

$$A = \begin{bmatrix} a & h \\ k & b \end{bmatrix}$$

of order $n + 1$, where a is of order n. By hypothesis, $vau = d$, where u, v and d are as stated. Also, by Schur's identity,

$$|A|/|a| = b - ka^{-1}h.$$

Then with

$$U = \begin{bmatrix} u & -a^{-1}u \\ o & 1 \end{bmatrix}, \quad V = \begin{bmatrix} v & o \\ -va^{-1} & 1 \end{bmatrix},$$

it follows that

$$VAU = \begin{bmatrix} d & o \\ o & b - ka^{-1}h \end{bmatrix}$$

so U, V are as required.

Corollary In the same way, there exist unit upper and lower triangular matrices u, v such that $a = vdu$.

For, by Theorem 2.2, if a matrix is unit upper or lower triangular, then so is its inverse.

9
Quadratic Forms

9.1 Symmetric matrices

Theorem 1.1 Any symmetric matrix has a regular principal submatrix from any base set of rows or columns.

Let a_{ij} be a critical submatrix of a, existing in any base set of rows i, or of columns j, so

$$a = a_{j)}\, a_{ij}^{-1}\, a_{(i}\,, \qquad (i)$$

where $a_{(i}\,, a_{j)}$ are the submatrices of such rows and columns. It has to be shown that the principal submatrices a_{ii} and a_{jj} are regular.
From (i),

$$a_{ji} = a_{jj}\, a_{ij}^{-1}\, a_{ii}\,, \qquad (ii)$$

But with a symmetric, $a_{ji} = (a_{ij})'$ is regular since a_{ij} is regular. Therefore from (ii), a_{ii} and a_{jj} must be regular. QED

Corollary (i) If a is a symmetric matrix, then it has a principal critical submatrix a_{ii} in any base set of rows or columns, and then

$$a = a_{i)}\, a_{ii}^{-1}\, a_{(i}\,.$$

For any regular submatrix of order equal to the rank of the matrix must be critical.

Corollary (ii) If a matrix is symmetric, then the principal submatrix on any independent set of columns is regular.

Replace by zero the elements not in the rows and columns of the principal submatrix. By the theorem, this submatrix is critical for the matrix so obtained.

Corollary (iii) If a symmetric matrix is regular, then so is every principal submatrix.

For its columns are independent, and hence so is any subset, and hence, by the last corollary, the principal submatrix from any subset of columns is regular.

The following comes from these theorems.

SYMMETRIC RANK REDUCTION The process of rank reduction of a symmetric matrix a always may proceed with pivots taken from the diagonal. On termination when the matrix has been reduced to o, the rank r together with a critical principal submatrix p will have been determined, and the sequence of pivots $p_1, \ldots, p_r$ provide successive ratios in a nested sequence of principal minors of that submatrix.

For the case where p is the leading principal submatrix of order r,

$$p_s = \delta_s/\delta_{s-1} \ (s = 1, \ldots, r),$$

where $\delta_0 = 1$, and δ_s is the leading principal minor of a of order s $(s = 1, \ldots, r)$. Then

$$p_1 \cdots p_s = \delta_s \ (s = 1, \ldots, r),$$

and

$$p_1 \cdots p_r = \delta_r = |p|.$$

This case, applicable in particular when a is regular, bears on the hypothesis in the theorem at the end of the last chapter, and will give further service in the present chapter.

9.2 Quadratic forms

A matrix being symmetric if $a' = a$ and antisymmetric if $a' = -a$. any matrix a is the sum $a = b + c$ of its *symmetric* and *antisymmetric parts* given by

$$b = \tfrac{1}{2}(a + a'), \ c = \tfrac{1}{2}(a - a').$$

Hence the *quadratic form* $x'ax$ with *coefficient matrix* a is a sum

$$x'ax = x'bx + x'cx.$$

But, with c antisymmetric,

$$x'cx = (x'cx)' = x'c'x'' = -x'cx,$$

so that $x'cx = 0$, showing that *if a matrix is antisymmetric, then the associated quadratic form is null.* We have

$$x'ax = x'bx,$$

that is, *a quadratic form is unchanged when the coefficient matrix is replaced by its symmetric part.* Hence there is no loss in generality in considering quadratic forms for which the matrix is symmetric.

Consider

$$f(x) = x'ax.$$

The gradient is

$$g(x) = x'(a + a'),$$

this being the row vector formed by the partial derivatives, and the Hessian, or matrix of second derivatives, is the constant symmetric matrix

$$h = a + a' = 2b,$$

where b is the symmetric part of a. It follows that

$$f(x) = \tfrac{1}{2} x'hx,$$

where h is the Hessian, and then

$$g(x) = x'h$$

is the gradient.

9.3 Definite forms

A matrix a is *non-negative definite,* or *positive semidefinite,* if $x'ax \geq 0$ for all x, and *positive definite* if $x'ax > 0$ for all $x \neq 0$. There are similar terms for these conditions applied to $-a$. We say a is *definite,* or *semidefinite,* if a or $-a$ has such properties, and otherwise a is *indefinite,* in which case $x'ax$ takes both positive and negative values.

Theorem 3.1 If a matrix is positive definite, then it is regular, and its inverse is positive definite.

For if a is singular, then $ax = o$ for some $x \neq o$. But then $x'ax = 0$, where $x \neq o$, so a is not positive definite.

Hence if a is positive definite, let b be the inverse which must now exist. For any $y \neq 0$, we have to show $y'by > 0$. Let $x = by$, so $x \neq o$, and $y = ax$. Then also

$$y'by = x'a'bax = x'ax,$$

so from $x'ax > 0$, by hypothesis since $x \neq o$, we have $y'by > 0$.
QED

Theorem 3.2 If b is of full column rank, then $a = b'b$ is positive definite.

For then
$$x'ax = x'b'bx = y'y \geq 0,$$

where $y = bx$ so, by hypothesis, $y \neq o$ if $x \neq o$. But in any case $y'y > 0$ if $y \neq o$. Hence $x'ax > 0$ if $x \neq o$. QED

9.4 Completing the square

Now consider a quadratic form $Q = z'cz$ where the variables z are partitioned into x's and y's. The coefficient matrix, correspondingly partitioned, has the form
$$c = \begin{bmatrix} a & u \\ v & b \end{bmatrix}$$

where a, b are symmetric, and $v = u'$. Then the quadratic form is
$$Q = [x' \quad y'] \begin{bmatrix} a & u \\ v & b \end{bmatrix} \begin{bmatrix} x \\ y \end{bmatrix} = x'ax + x'uy + y'vx + y'by.$$

If a is regular, with inverse a^{-1}, this may be put in the form
$$Q = (x' + y'va^{-1})\, a\, (x + a^{-1}uy) + y'(b - va^{-1}u)\, y.$$

The rearrangement generalizes the familiar 'completing the square' method for finding the roots of a quadratic
$$q = ax^2 + 2ux + b,$$

by putting it in the form
$$q = [(ax + u)^2 + (ab - u^2)]/a,$$

which shows the roots to be
$$x = \left(-u \pm \sqrt{u^2 - ab}\right)/a.$$

Theorem 4.1 Provided a is regular,
$$\begin{bmatrix} a & u \\ v & b \end{bmatrix} = \begin{bmatrix} 1 & o \\ va^{-1} & 1 \end{bmatrix} \begin{bmatrix} a & o \\ o & b - va^{-1}u \end{bmatrix} \begin{bmatrix} 1 & a^{-1}u \\ o & 1 \end{bmatrix}$$

This may be verified directly, and is another expression of the rearrangement given to Q. Alternatively, it may be obtained from Theorem 8.1.1.

Corollary (i)   For

$$\begin{bmatrix} a & u \\ v & b \end{bmatrix}$$

to be positive definite, it is necessary and sufficient that both of

$$a, \; b - va^{-1}u$$

be positive definite.

The necessity is from Theorem 3.1, and sufficiency is from the theorem above.

Theorem 4.2   A necessary and sufficient condition for a matrix to be positive definite is that its leading principal minors all be positive.

The proof is by induction on the order of the matrix. The case $n = 1$ is verified, and the inductive hypothesis is for the case for $n \geq 1$. Consider

$$c = \begin{bmatrix} a & u \\ v & b \end{bmatrix}$$

of order $n + 1$, where a is of order n. The *positive minor* condition which comes in the theorem, taken for c, is equivalent to the same on a together with $|c| > 0$, equivalently with $|c|/|a| > 0$ since the condition on a includes $|a| > 0$. Let

$$B = b - va^{-1}u,$$

so, given a is positive definite, $B > 0$ is equivalent to c being positive definite, by Corollary (i). Also, by Schur's identity,

$$|c|/|a| = B.$$

Hence, the positive minor condition on c is equivalent to that on a together with $B > 0$. By the hypothesis, this is equivalent to a being positive definite, together with $B > 0$. But, as just remarked, this is equivalent to c being positive definite. Hence the positive minor condition of c is equivalent to c being positive definite. QED

TEST ALGORITHM   This test should not be carried out in practice by directly evaluating the n minor determinants that are required to be positive. Instead one would simply perform a rank reduction procedure with pivots taken from the diagonal, and require that these pivots be n in number and all positive.

This comes from the observations on *symmetric rank reduction*, at the end of §1.

Theorem 4.3 For a symmetric matrix a of order n, with δ_m as its leading principal minor of order m, if

$$\delta_m > 0 \quad (m = 1, \ldots, n)$$

then there exists an upper triangular matrix b with diagonal elements given by

$$b_{mm} = \sqrt{\delta_m/\delta_{m-1}} \quad (m = 1, \ldots, n),$$

with $\delta_0 = 1$, such that

$$a = b'b.$$

This follows from Theorem 8.3.1 and Corollary, and, taken with Theorem 9.3.2, provides another proof of Theorem 4.2.

Corollary A matrix a is symmetric positive definite if and only if it is of the form $a = b'b$ where b is of full column rank.

This comes immediately from the theorem, and is the converse of Theorem 3.2.

9.5 Semidefinite forms

Theorem 5.1 If a regular symmetric matrix is positive semidefinite then it is positive definite.

For if a is regular and symmetric, then all its principal minors are non-zero, by Theorem 1.1, Corollary *(iii)*. Hence, by Theorem 8.3.1, Corollary, $a = u'du$, where d is a diagonal matrix whose elements must be non-negative if a is to be positive semidefinite. But they are non-zero, so they must be positive, making a positive definite.

Theorem 5.2 A symmetric matrix is positive semidefinite if and only if a principal critical submatrix is positive definite.

For

$$a = u'p^{-1}u$$

is positive semidefinite if and only if p^{-1}, equivalently p, is the same. But p is regular symmetric, so it must be positive definite.

Theorem 5.3 A necessary and sufficient condition for a symmetric matrix to be positive semidefinite is that all its principal minors be non-negative.

The proof is from the last two theorems. It is not sufficient for the sequence of leading principal minors to be non-negative, as shown by the example

$$a = \begin{bmatrix} 0 & 0 & 0 \\ 0 & 1 & 0 \\ 0 & 0 & -1 \end{bmatrix}.$$

TEST ALGORITHM This test should not be carried out in practice by directly evaluating the $2^n - 1$ minor determinants that are required to be non-negative. Instead one would simply perform a rank reduction procedure with pivots taken from the diagonal, and require these pivots, in number r equal to the rank, all to be positive.

As with the test algorithm in §4, this comes from the observations on symmetric rank reduction in §1,

Theorem 5.4 A matrix a is symmetric positive semidefinite if and only if it is of the form $a = b'b$, where a and b have the same rank.

For, if a is symmetric positive semidefinite, then $a = u'p^{-1}u$, where p^{-1} is positive definite and hence of the form $c'c$ where c is square and regular, so we have the required expression with $b = cu$. The converse is obvious.

Corollary If a is symmetric positive semidefinite, then

$$x'ax = 0 \iff ax = o.$$

9.6 Linear restrictions

Consider a quadratic form $x'ax$, where a is symmetric $n \times n$, and linear restrictions on the variables given by $ux = o$, where u is

$m \times n$. The quadratic form is positive definite subject to the restrictions if

$$ux = o, \ x \neq o \ \Rightarrow \ x'ax > 0.$$

The obvious, and the original, method for dealing with a quadratic subject to constraints is to eliminate the constraints.

Let the constraint matrix be partitioned as $u = [p \ \ r]$, where p is square and regular. Correspondingly, the symmetric coefficient matrix and the variables are partitioned as

$$a = \begin{bmatrix} b & h \\ k & d \end{bmatrix}, \ x = \begin{bmatrix} y \\ z \end{bmatrix}$$

where b, d are symmetric and $k = h'$. Now

$$ux = py + rz$$

so $ux = o$ is equivalent to

$$y = -p^{-1}rz,$$

and hence to

$$x = \begin{bmatrix} -e \\ 1 \end{bmatrix} z,$$

where $e = p^{-1}r$. With

$$H = [-f \ \ 1] \begin{bmatrix} b & h \\ k & d \end{bmatrix} \begin{bmatrix} -e \\ 1 \end{bmatrix} = fbe - fh - ke + d.$$

where $f = e'$, we have $x'ax$ is positive definite, or semidefinite, subject to $ux = o$ if and only if H is unrestrictedly positive definite, or semidefinite.

Conditions for H to be positive definite are expressed in terms of its leading principal minors. It will be seen how these have expression in terms of leading principal minors of the bordered matrix

$$A = \begin{bmatrix} o & u \\ u' & a \end{bmatrix}.$$

It follows from Theorem 8.1.2, Corollary, that

$$|A| = (-1)^m |p|^2 |H|.$$

For any set i of indices $i_1, i_2, \ldots$ where

$$1 \leq i_1 < i_2 < \cdots \leq n - m,$$

let H_i denote the principal submatrix of H drawn on the rows and columns with these indices, and let

$$\Delta_i = (-1)^m \begin{vmatrix} o & u_i \\ u_i' & a_i \end{vmatrix} / |p|^2,$$

where u_i is the submatrix of u on the columns $1, \ldots, m$ and then the further columns $m + i$ for $i \in i$, and a_i is defined similarly. Hence

$$a = a_n, \; u = u_n.$$

Theorem 8.1.2, Corollary, shows that

$$|H_i| = \Delta_i \,,$$

and conditions for H to be positive definite, or semidefinite, can now be stated with this expression for any principal minor.

Theorem 6.1 [12] A necessary and sufficient condition for $x'ax$ to be positive definite subject to $ux = o$ is that

$$(-1)^m \begin{vmatrix} o & u_i \\ u_i' & a_i \end{vmatrix} > 0 \;\; (r = m + 1, \ldots, n),$$

provided $|u_m| \neq 0$, where a_r is the leading $r \times r$ submatrix of a, and u_r is the leading $m \times r$ submatrix of u.

This corresponds to Theorem 4.2 for an unrestricted form, and Theorem 5.3 likewise produces a result for the semidefinite case. As there, it suffices to find a critical principal submatrix, and the sign conditions can apply just to this and its principal minors, hence the following.

Theorem 6.2 If

$$A = \begin{bmatrix} o & u \\ u' & a \end{bmatrix}, \;\; B = \begin{bmatrix} o & v \\ v' & b \end{bmatrix}$$

are such that A is symmetric and has B as a principal critical submatrix, then

$$ux = o \;\Rightarrow\; x'ax \geq 0$$
$$\Updownarrow$$
$$vy = o, \, y \neq o \;\Rightarrow\; y'bx > 0.$$

Also,

TEST ALGORITHM In the course of a symmetric rank reduction procedure, by which B would be discovered from A, the pivots, being ratios of minors that are required to have the same sign, must all be positive.

---

[12] W. L. Ferrar (1951) deals with the case $m = 1$ and remarks about a generalization which he attributes to myself. Aware of this case, he had suggested that I attempt the generalization published in my paper of 1950. As now appears, the theorem had already been produced by H. B. Mann (1943).

Remarks now will concern conditions for a quadratic form to be definite on a subspace expressed in terms of a quadratic form in the Grassmann coordinates, or the dual coordinates. With

$$t = \begin{bmatrix} -e \\ I \end{bmatrix},$$

so the columns of t are a solution base for $ux = o$, we have

$$H = t'at.$$

Then, since

$$\begin{vmatrix} a^{-1} & t \\ t' & o \end{vmatrix} = |a|^{-1} |-t'at|,$$

it follows that

$$|H| = (-1)^m |a| \begin{vmatrix} a^{-1} & t \\ t' & o \end{vmatrix}.$$

Also,

$$|A| = (-1)^m |a| |ua^{-1}u'|,$$

and so, from Theorem 1.2, Corollary,

$$|a| |ua^{-1}u'| = |p|^2 |t'at|.$$

For the right, by the Binet-Cauchy theorem,

$$|t'at| = T'a^{(n-m)} T.$$

where $T = t^{(n-m)}$ is the vector of Grassman coordinates for the column space of t. Similarly for the left,

$$|ua^{-1}u'| = U'(a^{(m)})^{-1}U = U'a^{[m]} U/|a|,$$

where $U = u^{(m)'}$ is the vector of dual Grassmann coordinates for the same space. Then

$$T'a^{(n-m)}T = U'a^{[m]}U/|p|^2,$$

which shows the connection between statements given in the coordinates and in the dual coordinates.

9.7 Finsler's theorem

Theorem 7.1 If $x'ax$ is positive definite subject to $ux = o$, and ρ^* is the least root of the equation

$$|a - u'\rho u| = 0,$$

then $x'(a - u'\rho u)x$ is positive definite for all $\rho < \rho^*$.

Under the hypothesis, the function

$$\rho(x) = x'ax/x'u'ux,$$

regarded as a function to the extended real numbers with their usual topology, is well defined, and such that $\rho(x) > -\infty$, for $x \neq o$, and has gradient

$$g(x) = 2(x'u'ux)^{-1}x'(a - u'\rho(x)u).$$

It is continuous, and homogeneous of degree zero, so all its values are taken when x is restricted to the unit sphere. Hence, by compactness, it attains a minimum ρ^* at some $x^* \neq o$. Then $ux^* \neq o$ and $g(x^*) = o$, so that

$$(a - u'\rho^*u)x^* = o,$$

and hence

$$|a - u'\rho^*u| = 0.$$

But all the roots of this equation are real, and all are values of $\rho(x)$. Since ρ^* is the minimum value it must therefore be the minimum root. Hence we have

$$\rho(x) \geq \rho^* \text{ for all } x \neq o,$$

so that, if $\rho < \rho^*$, then

$$\rho(x) > \rho \text{ for all } x \neq o.$$

QED

It can be added that $a - u'\rho^*u$ is positive semidefinite, and so

$$x'(a - u'\rho^*u)x = 0 \Leftrightarrow (a - u'\rho^*u)x = o.$$

Also, the nullity of $a - u'\rho^*u$, which is the dimension of its null space, is equal to the multiplicity of ρ^* as a root of the equation

$$|a - u'\rho u| = 0.$$

Corollary (i) A necessary and sufficient condition for $x'ax$ to be positive definite subject to $ux = o$ is that $x'(a - u'\rho u)x$ be positive definite for some ρ.

The necessity is shown by the theorem, and the sufficiency is obvious.

If the condition holds for some ρ then it must hold for all $\rho' \leq \rho$. Hence we have

Corollary (ii) A necessary and sufficient condition for $x'ax$ to be positive definite subject to $ux = o$ is that

$$x'(a + u'\lambda u)x$$

be positive definite for sufficiently large λ.

This is Finsler's Theorem.

Corollary (iii)   A necessary and sufficient condition for $x'ax$ to be positive definite is that it be positive definite subject to $ux = o$ and the roots of

$$|a - u'\rho u| = 0$$

all be positive.

For if $x'(a - u'\rho u)x$ is positive definite where $\rho > 0$, then $x'ax$ is positive definite.

Consider a symmetrically partitioned symmetric matrix

$$c = \begin{bmatrix} a & h \\ h' & b \end{bmatrix}$$

where a is regular. With variables

$$z = \begin{bmatrix} x \\ y \end{bmatrix}, \quad w = [o \ 1],$$

the constraint $wz = o$ is equivalent to $y = o$. Hence $z'cz$ being positive definite subject to $wz = o$ is equivalent to $x'ax$ being positive definite.

It has already been seen that for c to be positive definite it is necessary and sufficient that both of

$$a, \ b - h'a^{-1}h$$

be positive definite. Now we will be able to see Corollary (iii) as a generalization of this result. For it will be found that the roots of the equation

$$|c - w'\rho w| = 0$$

are identical with the characteristic roots of the matrix

$$b - h'a^{-1}h.$$

Then these being all positive is exactly the condition for $b - h'a^{-1}h$ to have its characteristic roots all positive, and hence for it to be positive definite.

By Schur's identity,

$$|c - w'\rho w| = \begin{vmatrix} a & h \\ h' & b - \rho 1 \end{vmatrix} = |a||b - \rho 1 - h'a^{-1}h|.$$

For the case $m = 1$, and $|a| \neq 0$, by two applications of Schur's identity

$$\rho^{-1}|a - u'\rho u| = \begin{vmatrix} \rho^{-1} & u \\ u' & a \end{vmatrix} = |a|(\rho^{-1} - ua^{-1}u'),$$

which shows that

$$\rho^* = (ua^{-1}u')^{-1}$$

is the unique root of

$$|a - u'\rho u| = 0.$$

Corollary (iv)[13] A necessary and sufficient for $x'ax$ to be positive semidefinite under a single linear restriction $ux = 0$ is that

$$a - u'(ua^{-1}u')u$$

be positive semidefinite, and given this, a necessary and sufficient condition for a to be positive semidefinite is that

$$ua^{-1}u' \geq 0.$$

For another approach, if u is of full rank $m \leq n$, then uu' is regular, and the symmetric idempotent

$$I_u = u'(uu')^{-1}u \quad [14]$$

is the orthogonal projector onto the row space of u, and

$$J_u = I - I_u$$

is the orthogonal projector onto the orthogonal complement, which is the solution space of $ux = o$. Then the values of

$$x'J_u a J_u x$$

without restriction, are identical with the values of $x'ax$ subject to $ux = 0$. It follows that

$$ux = o \Rightarrow x'ax \geq 0 . \Leftrightarrow . x'J_u a J_u x \geq 0 \text{ for all } x.$$

It can be added, $x'ax$ is positive definite under $ux = o$ if and only if $J_u a J_u$, which in any case has zero as a characteristic value of multiplicity at least m, has positive characteristic values of total multiplicity $n - m$, in which case, necessarily, zero has multiplicity

---

[13] This result has application to the convexification of quasiconvex functions dealt with by Bruno de Finetti (1949) and W. Fenchel (1953). An account is in my 1980 book, Chapter V. It is also related to the breakdown of the convexity of a function into the convexity of its contours and its profiles, as shown in Chapter IV.

[14] This formula, which originates in my 1957b paper in the general form and various extensions, has many applications, such as in multivariate analysis, or gradiant projection methods.

exactly m. An equivalent condition is that $J_u a J_u$ be positive semidefinite and moreover of rank $n - m$.

By Lagrange's method, it appears that the characteristic values of $J_u a J_u$, which, since J_u is idempotent, are the same as the characteristic values of $a J_u J_u = a J_u$, are the stationary values of $x'ax$ subject to $ux = o$, $x'x = 1$. The minimum exists, by compactness of the constraint set, and is the least of these values, so the considered condition is just that these values all be positive, as already established by another route. The Lagrange conditions are

$$2x'a - \lambda u - 2\mu x' = o,$$

together with

$$ux = o, \quad x'x = 1.$$

From these,

$$x'ax = \mu, \quad 2x'au' = \lambda uu',$$

so that

$$\lambda = 2x'au'(uu')^{-1}, \quad \lambda u = 2x'aJ_u.$$

Then

$$x'(aJ_u - \mu I) = o, \quad x \neq o,$$

and hence

$$|aJ_u - \mu I| = 0,$$

which provides the conclusion just stated.

Let a be $n \times n$, and u be $m \times n$ of rank m. Without loss in generality, it can be assumed that the leading m columns of u are independent. For $r = 1, \ldots, n$, let a_r denote the leading $r \times r$ submatrix of a, and u_r the leading $m \times r$ submatrix of u, so

$$a = a_n, \quad u = u_n.$$

By assumption, u_m is a regular $m \times m$ matrix,

$$|u_m| \neq 0.$$

This implies

$$|u'_r u_r| > 0 \quad (r = 1, \ldots, m).$$

With

$$D_r(\rho) = |a_r - u'_r \rho u_r|,$$

the condition for $a - u'\rho u$ to be positive definite is that

$$D_r(\rho) > 0 \quad (r = 1, \ldots, n).$$

The condition for $x'ax$ to be positive definite subject to $ux = o$ is that this holds for some ρ, or equivalently, when $\rho \to -\infty$.

But since $|u'_r u_r| > 0$, the sign of $D_r(\rho)$ agrees with that of

$$|-u_r' \rho u_r| = (-\rho)^r |u_r' u_r|,$$

for large $|\rho|$. Thus in any case

$$D_r(\rho) > 0 \text{ as } \rho \to -\infty \ (r = 1, \dots, m),$$

so it remains to consider $D_r(\rho)$ for $r = m + 1, \dots, n$.

By Schur's identity,

$$\begin{vmatrix} 1\rho^{-1} & u_r \\ u_r' & a_r \end{vmatrix} = |1\rho^{-1}||a_r - u_r' \rho u_r| = \rho^{-m} D_r(\rho).$$

Let

$$\Delta_r = (-1)^m \begin{vmatrix} o & u_r \\ u_r' & a_r \end{vmatrix}.$$

Now P can denote the condition that $x'ax$ be positive definite subject to $ux = o$, and Δ the condition

$$\Delta_r > 0 \ (r = m + 1, \dots, n).$$

Also let N be the condition

$$\Delta_r \neq 0 \ (r = m + 1, \dots, n),$$

so immediately,

$$\Delta \Rightarrow N.$$

It will be seen that

$$P \Rightarrow N.$$

Suppose $\Delta_r = 0$ for some $r > m$, contrary to N. Then the equations

$$a_r x_r + u_r' y_m = o$$

$$u_r x_r + o \quad = o$$

have a solution $(x_r, y_m) \neq o$. Then $x_r \neq o$. For otherwise $u_r' y_m = o$, and since $|u_m| \neq 0$, and $r > m$, this implies also $y_m = o$. Also,

$$0 = x_r' a_r x_r + g x_r' u_r' y_m = x_r' a_r x_r,$$

so P is denied, giving the required conclusion.

We now have

$$(P \vee \Delta) \Rightarrow N,$$

and it will be shown that

$$N \Rightarrow (P \Leftrightarrow \Delta),$$

so it will follow that

$$P \Leftrightarrow \Delta.$$

It has been seen that P is equivalent to

$$D_r(\rho) = \rho^m \begin{vmatrix} 1\rho^{-1} & u_r \\ u_r' & a_r \end{vmatrix} > 0 \ (r = m + 1, \dots, n)$$

as $r \to -\infty$. But, given N, so that $\Delta_r \neq 0$, this, by continuity, is equivalent to $\Delta_r > 0$, that is the condition Δ, as required. Hence the following, which reproduces Theorem 6.1.

Corollary (**v**) A necessary and sufficient condition for $x'ax$ to be positive definite subject to $ux = o$ is that

$$(-1)^m \begin{vmatrix} o & u_r \\ u_r' & a_r \end{vmatrix} > 0 \quad (r = m+1, \ldots, n),$$

provided $|u_m| \neq 0$, where a_r is the leading $r \times r$ – submatrix of a, and u_r is the leading $m \times r$ submatrix of u.

Appendix

1 Permutations

A *permutation* of $1, \ldots, n$ is a mapping π of the set into itself such that

$$i \neq j \;\Rightarrow\; \pi i \neq \pi j,$$

that is, distinct elements have distinct images, so the mapping is one-one. The number of images being also n, this must be a one-one mapping onto the set, and hence with inverse π^{-1}, which must also be a permutation of the set. Let P be the set of these permutations. The *identity* permutation $1 \in P$ is defined by $1i = i$. The *product* of any ρ, $\sigma \in P$ is the mapping $\rho\sigma$ defined by

$$(\rho\sigma)i = \rho(\sigma i).$$

Then $\rho\sigma \in P$, since

$$i \neq j \;\Rightarrow\; \sigma i \neq \sigma j \;\Rightarrow\; \rho(\sigma i) \neq \rho(\sigma j) \;\Rightarrow\; (\rho\sigma)i \neq (\rho\sigma)j.$$

The product operation is associative, $\rho(\sigma\tau) = (\rho\sigma)\tau$, since

$$(\rho(\sigma\tau))i = \rho((\sigma\tau)i) = \rho((\sigma(\tau i))) = (\rho\sigma)(\tau i) = ((\rho\sigma)\tau)i.$$

Hence *powers* π^r ($r = 1, 2, \ldots$) defined by

$$\pi^0 = 1, \quad \pi^r = \pi\pi^{r-1}$$

have the property

$$\pi^r \pi^s = \pi^{r+s} = \pi^s \pi^r.$$

Also,

$$1\pi = \pi = \pi 1,$$

and

$$\pi\pi^{-1} = 1 = \pi^{-1}\pi,$$

for all $\pi \in P$. Hence, with this multiplication, P is a group. As the group of permutations on n objects, it defines the *symmetric group* of *degree n*.

The elements *affected* by π are those i for which $\pi i \neq i$. Though in general $\rho\sigma \neq \sigma\rho$, if the sets of elements affected by two permutations ρ, σ are disjoint, then $\rho\sigma = \sigma\rho$.

Apart from the identity permutation, which affects no elements, the simplest case of a permutation is the *transposition*, which affects just two elements. Thus for a transposition $\tau \in P$, on elements i, j

$$\tau i = j, \quad \tau j = i, \quad \tau k = k \quad (k \neq i, j).$$

This permutation may be denoted $\tau = (ij)$.

A further simple case, of which the transposition is an example, is the *cyclic permutation*, which affects a subset of elements, mapping each into its successor in a cyclic order. For cyclic $\gamma \in P$ affecting $i, j, \ldots, k$ in this cyclic order, where the first follows the last, we have

$$\gamma i = j, \ldots, \gamma k = i,$$

while

$$\gamma h = h \quad (h \neq i, j, \ldots, k).$$

Such a permutation can be denoted $\gamma = (ij \ldots k)$. It should be understood that the elements $i, j, \ldots, k$ are all be distinct; that is, this sequence should be a *simple* cycle. Otherwise the cyclic sequence must consist of a repetition of some simple cycle, which alone would define the same cyclic permutation. Now it can be asserted that

$$\gamma = (i\, \gamma i \ldots \gamma^{m-1}i), \quad \gamma^m i = i,$$

where m is the number of distinct elements, making this permutation a *cycle* of *order m*, or an *m-cycle*. Similarly, i could be replaced by any of the other elements $j, \ldots, k$ that are affected by γ, to provide another such representation of c.

Now given any permutation $\pi \in P$, and element i, there will be a smallest m from $m = 1, 2, \ldots$ such that $\pi^m i = i$, this m being the *order* of i in π. Then we can form the cyclic permutation

$$\gamma_i = (i\, \pi i \ldots \pi^{m-1}i),$$

this being the cycle of π on i, the case where $m = 1$ and $\gamma_i = (i)$ being admitted. Elements i, j are *cocyclic* in π if $\gamma_i = \gamma_j$. This defines an equivalence relation between elements, partitioning them into classes, each associated with a cycle of π. We now have a set of cycles of p where the elements affected form a partition of all the elements, and π is the product of these cycles, taken in any order, since the sets of affected elements are disjoint so they all commute. The following has been proved.

Theorem 1 Every permutation is a product of permutations on cocyclic classes that form a partition of the elements.

Provided

$$\pi \neq (1) \cdots (n) = 1,$$

we may omit those cycles that are of order 1, or are on a single element, and still have such an expression for π. With disjoint cycles the sets of elements affected are disjoint. Now we have the following.

Corollary Every permutation different from the identity is a product of disjoint cycles.

From the next theorem it will be possible to conclude that every permutation is a product of transpositions, if we accept a cycle of order 1 as a transposition, or exclude the identity permutation.

Theorem 2 Every cyclic permutation is a product of transpositions.

If the order is 1, or we exclude this case, and even if the order is 2, which case is immediately verified, there is nothing more to prove. Hence assume an order $m > 2$. Following Halmos (1958, p. 45) more or less, we may now submit that

$$(ij \ldots k) = (ik) \ldots (ij),$$

and this constitutes the proof.

If we do not like to accept a transposition of order 1 as a proper transposition, we will require the permutation in the following to be different from the identity, or to be a proper permutation.

Theorem 3 Every permutation is a product of transpositions.

This comes from the last two theorems. One can also proceed more directly, by induction on the number of elements affected in any permutation. Let

$$(i' \ldots k' \mid i \ldots k)$$

stand for the permutation π which makes

$$\pi i = i', \ldots, \pi k = k',$$

so any

$$\pi = (\pi 1 \ldots \pi n \mid 1 \ldots n),$$

and $(ij \mid ji)$ is the transposition (ij). Then

$$\pi' = (\pi 1\ 1 \mid 1\ \pi^{-1}1)\pi$$
$$= (1\ \pi'2 \ldots \pi'n \mid 1\ 2 \ldots n)$$

$$= (\pi'2 \ldots \pi'n \mid 2 \ldots n).$$

We can now apply the inductive hypothesis to π', to obtain transpositions $\tau_2, \ldots, \tau_n$ to make

$$\tau_n \ldots \tau_2 \pi' = 1.$$

The inverse of a transposition being a transposition, with

$$\tau_1 = (\pi 1 \ 1 \mid 1 \ \pi^{-1}1)$$

we then have

$$\pi = \tau_1 \ldots \tau_n 1.$$

Of course some and even all of these transpositions may have order 1, and be left out.

The elements $1, -1$ form a group A with multiplication, and we are going to consider a mapping

$$\chi : P \rightarrow A,$$

of the permutation group P onto A, which is a homomorphism, that is,

$$\chi(\rho\sigma) = (\chi\rho)(\chi\sigma),$$

where $\chi\tau = -1$ for any transposition τ, and $\chi 1 = 1$. Such a homomorphism is an *alternating character* for the permutation group, and also for any subgroup affecting a subset of elements.

   With a permutation π expressed as a product of m transpositions, it would be concluded that $\chi\pi = (-1)^m$. There are many such expressions for any permutation. However it would follow, from the existence of the alternating character, that the number m of factors must be always odd or always even for a given permutation. Accordingly, permutations can be classified for their *parity*, determined as *even* or *odd*, and this would match the alternating character being 1 or -1.

   For the existence of an alternating character, consider a function of n variable $x = (x_1, \ldots, x_n)$ given by

$$f(x) = \prod_{i<j} (x_i - x_j).$$

If the variables are permuted, this can only alter the sign of f. Moreover, with

$$\pi x = (x_{\pi 1}, \ldots, x_{\pi n}),$$

so

$$(\rho\sigma)x = \rho(\sigma x),$$

we have

$$f(\pi x) = \alpha(\pi) f(x),$$

where $\alpha(\pi) = 1$ or -1, and

$$\alpha(\rho\sigma)f(x) = f((\rho\sigma)x) = f(\rho(\sigma x))$$
$$= \alpha(\rho)f(\sigma x) = \alpha(\rho)\alpha(\sigma)f(x),$$

so that

$$\alpha(\rho\sigma) = \alpha(\rho)\alpha(\sigma).$$

Also,

$$f(\tau x) = -f(x)$$

if τ is a transposition. Hence α is an alternating character for the permutations. Uniqueness being assured by the value on transpositions, together with Theorem 7.3, we can take $\chi = \alpha$.

With the unit matrix $I \in K_n^n$, and the permutation matrix I_π as this after the columns have received a permutation π, we have

$$I_\rho I_\sigma = I_{\sigma\rho},$$

for all $\rho, \sigma \in P$, so the permutation matrices are close under multiplication. They form a group (anti)isomorphic with P under the mapping

$$\pi \rightarrow I_\pi.$$

From the definition of the determinant,

$$|I_\pi| = \chi(\pi)|I| = \chi(\pi).$$

Also, in consistency with the properties of χ, the multiplication theorem for determinants shows that

$$\chi(\rho\sigma) = |I_{\rho\sigma}| = |I_\sigma I_\rho| = |I_\sigma||I_\rho| = \chi(\sigma)\chi(\rho).$$

2 Combinations

Listing submatrices of a matrix requires a listing of combinations, or subseries of the series of rows and columns. For instance, with the derived matrices of the Binet-Cauchy theorem, the elements are subdeterminants. Every given subseries should determine a location index, and from any index one should be able to construct the subseries that belongs to it.

A subseries of k elements of the series $1, \ldots, n$ $(k \leq n)$ is given by

$$i = (i_1, \ldots, i_k), \quad 1 \leq i_1 < \cdots < i_k \leq n.$$

The number of such series is the number of ways k elements can be chosen from n, given by the *binomial coefficient*

$$\binom{n}{k} = \frac{n!}{k!\,(n-k)!}$$

where

$$n! = n(n-1)\cdots 1$$

is the *factorial* of n, with $0! = 1$.

The subseries can be ordered by the *order of last differences*, in which priority is decided by that of their last differing elements. With

$$\boldsymbol{i} = (i_1,\dots,i_k),\quad \boldsymbol{j} = (j_1,\dots,j_k),$$

let the last element in which they differ be their rth, so that

$$i_r \neq j_r,\quad i_s = j_s\ (s > r).$$

Then $\boldsymbol{i} < \boldsymbol{j}$ or $\boldsymbol{j} < \boldsymbol{i}$ according as $i_r < j_r$ or $j_r < i_r$. This relation $<$ between subseries is evidently a total order, forming them into a series.

Now let $n^{(k)}$ $(1 \le k \le n)$ be the series of subseries of k elements of the series $(1,\dots,n)$ with the order of last differences, or the kth *derived series*. In particular,

$$n^{(1)} = ((1),\dots,(n)),\quad n^{(n)} = ((1,\dots,n)),$$

and generally,

$$n^{(k)} = ((1,\dots,k),\dots,(n-k+1,\dots,n)).$$

Let $n_i^{(k)}$ denote the subseries which is the ith term of $n^{(k)}$.

Theorem 1 For $k \le m \le n$, $m^{(k)}$ is the subseries of $n^{(k)}$ of the leading $\binom{m}{k}$ terms.

Firstly, $m^{(k)}$ is a subseries of $n^{(k)}$ of $\binom{m}{k}$ terms. Also, any term of $m^{(k)}$ which does not belong to $m^{(k)}$ has its final element exceeding m, and so coming after all the elements of every term of $m^{(k)}$.

Lemma $\binom{n-1}{k} + \binom{n-1}{k-1} = \binom{n}{k}$.

This can be seen either directly, by cancelling factors so it reduces to

$$\frac{1}{k} + \frac{1}{n-k} = \frac{n}{k\,(n-k)},$$

or by applying the binomial theorem on either side of

$$(1+x)^n = (1+x)(1+x)^{n-1},$$

and equating coefficients.

Corollary $\sum_{k \le m \le n} \binom{m-1}{k-1} = \binom{m}{k}$.

Theorem 2 For given $k \le n$, and

$$1 \le i_1 < \ldots < i_k \le n, \ 1 \le s \le n,$$

the subseries I_s of $n^{(k)}$ of terms of the form

$$(x_1, \ldots, x_s, i_{s+1}, \ldots, i_k)$$

is a consecutive subseries of length

$$\binom{i_{s+1} - 1}{s}$$

following the N_sth term in $n^{(k)}$, where

$$N_{s-1} - N_s = \binom{i_s - 1}{s}.$$

The terms of I_s are in correspondence with the terms

$$(x_1, \ldots, x_s)$$

of the series $(i_{s+1} - 1)^{(s)}$ of length

$$\binom{i_{s+1} - 1}{s}.$$

The last element in which any term σ of I_s differs from any other term ρ of $n^{(k)}$ must be the t-th where $t > s + 1$, since $t \le s + 1$ would make ρ belong to I_s. The other terms of $n^{(k)}$ therefore have the form

$$\rho = (y_1, \ldots, y_s, y_{s+1}, \ldots, y_t, i_{t+1}, \ldots, i_k),$$

where $t > s + 1$ and $y_t \ne i_t$, while any term of I_s is

$$(x_1, \ldots, x_s, i_{s+1}, \ldots, i_t, i_{t+1}, \ldots, i_k).$$

Then ρ comes before or after any σ according as $y_t <$ or $> i_t$. It follows that the terms of I_s are consecutive, since any other term of $n^{(k)}$ either before or after all of them.

Consider two subseries of I_s, one consisting of terms with $x_s < i_s$ and the other of terms with $x_s = i_s$. By Theorem 1 they both are leading consecutive subseries in I_s, and the first has

$$\binom{i_s - 1}{s}$$

terms. Therefore the second, which is identical with I_{s-1}, is a series of consecutive terms following the

$$\binom{i_s - 1}{s} \text{-th term of } I_s,$$

and the

$$N_{s-1}\text{-th term of } n^{(k)}.$$

Since I_s follows the N_s-th term of $n^{(k)}$, this shows the required difference relation for N_s.

Corollary (i) $N_s = \sum_{s+1 \leq r \leq k} \binom{i_r - 1}{r}$.

Since $N_k = 0$, because I_k is identical with the entire series $n^{(k)}$, this follows by summation of the difference relation for N_s provided by the theorem.

Corollary (ii) If $n_i^{(m)} = (i_1, \ldots, i_k)$ then
$$i = 1 + \sum_{1 \leq s \leq k} \binom{i_s - 1}{s}.$$

The series I_0 consists in the one term $n_i^{(k)}$. This being the ith term of $n^{(k)}$ and also the term of $n^{(k)}$ after the N_0-th, it follows that $i = 1 + N_0$ and, with the formula for N_0 provided by Corollary (i), hence the required conclusion.

Corollary (iii) If $n_i^{(k)} = (i_1, \ldots, i_k)$, then i_s is determined from i and $i_{s+1}, \ldots, i_k$ by the condition
$$N_s + \binom{i_s - 1}{s} < i \leq N_s + \binom{i_s}{s}$$

By the theorem, I_s is a consecutive series of
$$\binom{i_{s+1} - 1}{s}$$
terms following the N_sth in $n^{(k)}$; that is, the positions i in $n^{(k)}$ of the terms of I_s are those which satisfy
$$N_s < i \leq N_s + \binom{i_{s+1}-1}{s}, \tag{i}$$
where N_s, given by Corollary (i), depends only on $i_{s+1}, \ldots, i_k$. The hypothesis for the present corollary requires this condition for $s = k, k-1, \ldots, 1$. Replacing s in (i) by $s - 1$ and using the difference relation for N_s,
$$N_s + \binom{i_s-1}{s} < i \leq N_s + \binom{i_s-1}{s} + \binom{i_s-1}{s-1}. \tag{ii}$$
By the Lemma, this is the relation this corollary requires. It remains to see that, given $i_{s+1}, \ldots, i_k$ and i which satisfy (i), there exists just one i_s which satisfies (ii).

The series of segments
$$J_{sm} = \left\{ x : N_s + \binom{m-1}{s} < x \leq N_s + \binom{m}{s} \right\}$$

for $m = s, s+1, \ldots, i_{s+1} - 1$ make a partition of the segment

$$J_s = \left\{ x : N_s < x \leq N_s + \binom{i_{s+1}-1}{s} \right\}.$$

By hypothesis i belongs to J_s, as stated by (i). Consequently i belongs to just one of the segments J_{sm}, determining $i_s = m$ so that (ii) is satisfied.

Corollary (iv) For given n, $k \leq n$ and $1 \leq i \leq \binom{n}{k}$, there exist unique $i_1, \ldots, i_k$ determined successively so that when $i_{s+1}, \ldots, i_k$ have been determined, i_s is then determined so that

$$N_s + \binom{i_s-1}{s} < i \leq N_s + \binom{i_s}{s},$$

where

$$N_{s-1} - N_s = \binom{i_s-1}{s}.$$

The $i_1, \ldots, i_k$ so determined are such that

$$i = 1 + \sum_{1 \leq s \leq k} \binom{i_{s-1}-1}{s}.$$

and

$$n_i^{(k)} = (i_1, \ldots, i_k).$$

For every such i there exists a series $n_i^{(k)}$ which has the ith position in $n^{(k)}$ and, by Corollary (ii), it can be constructed in the stated way.

Corollary (v) For any n and $k \leq n$,

$$1 + \sum_{1 \leq r \leq n} \binom{n-k+r+1}{r} = \binom{n}{k}.$$

For the series with $i_r = n - k + r$ is the last in $n^{(k)}$, with position $i = \binom{n}{k}$, so this follows incidentally from Corollary (ii).

Any subseries of k elements of $(1, \ldots, n)$ has a complementary subseries of $n - k$ elements. Let $n_i^{[k]}$ be the complement of $n_i^{(k)}$, this being the ith term of the kth *complementary derived series* $n^{[k]}$ of $(1, \ldots, n)$. This is the series of the

$$\binom{n-k}{k} = \binom{n}{k}$$

subseries of $n - k$ elements taken in reverse order of last differences.

Theorem 3 The series $n^{[k]}$ is the series $n^{(n-k)}$ taken in its reverse order.

To see this, let

$$i = (i_1, \ldots, i_k), \ j = (j_1, \ldots, j_k)$$

have complements

$$i' = (i'_1, \ldots, i'_{n-k}), \ j' = (j'_1, \ldots, j'_{n-k}).$$

It has to be shown that if i precedes j in the order of last differences, then i' succeeds j'. Let the last elements in which i, j differ be their rth, so that

$$i = (i_1, \ldots, i_r, x_{r+1}, \ldots, x_k),$$
$$j = (j_1, \ldots, j_r, x_{r+1}, \ldots, x_k),$$

where

$$i_r < j_r < x_{k+1}.$$

Now j_r occurs in i', as the sth element say, so that $i'_s = j_r$, and this will be the last element in which i' differs from j'. Further, $j'_s < j_r$, so now $j'_s < i'_s$, showing that j' precedes i'.

The complementary subseries $n_i^{(k)}$ and $n_i^{[k]}$ adjoin to give a series

$$(n_i^{(k)}, n_i^{[k]})$$

which is a derangement of $(1, \ldots, n)$, by a permutation

$$\pi_i = (n_i^{(k)}, n_i^{[k]})/(1, \ldots, n).$$

With

$$n_i^{(k)} = (i_1, \ldots, i_k), \ n_i^{[k]} = (i'_1, \ldots, i'_{n-k})$$

this permutation is such that

$$\pi_i r = i_r \ (r \le k), \ \pi_i r = i'_{k-r} \quad (r > k).$$

Let

$$\chi_i = \chi(\pi_i)$$

denote the alternating character of this permutation, ± 1 according as it is an even or odd permutation, that is, as the number of factors in any decomposition of it into a product of transpositions is even or odd.

Theorem 4 If $n_i^{(k)} = (i_1, \ldots, i_k)$, then

$$\chi_i = (-1)^{\sum_{1 \le r \le k} i_r - \frac{1}{2}k(k+1)}.$$

By $i_1 - 1$ transpositions, each of which exchanges i_1 with an element of $n_i^{[k]}$, and so preserves the order of the elements in $n_i^{[k]}$, i_1 in

$$(1, \ldots, i_1, \ldots, i_k, \ldots, n)$$

can be brought to position 1. Then similarly, $i_2 - 2$ transpositions bring i_2 to position 2, and so forth, until finally

$$\sum_{1 \leq r \leq k}(i_r - r) = \sum_{1 \leq r \leq k}i_r - \tfrac{1}{2}k(k+1)$$

transpositions have brought $i_1, \ldots, i_k$ to positions $1, \ldots, k$ while the remaining elements, those in the complement $n_i^{[k]}$, are in the order they had originally, so the result is

$$(n_i^{(k)}, n_i^{[k]}),$$

giving the required conclusion.

Corollary (i)   If $n_i^{[k]} = (i'_1, \ldots, i'_{n-k})$, then

$$\chi_i = (-1)^{\frac{1}{2}(n-k)(n+k+1) - \sum_{1 \leq s \leq n-k} i'_s}.$$

For

$$\sum_{1 \leq r \leq k} i_r + \sum_{1 \leq s \leq n-k} i'_s = \sum_{1 \leq m \leq n} m = \tfrac{1}{2}n(n+1).$$

Let

$$\chi_{ij} = \chi_i \chi_j,$$

this being the sign which, for the minor of a determinant taken from the ith set of k rows and jth set of k columns, must be attached to the complementary minor to make the cofactor.

Corollary (ii)   If

$$n_i^{(k)} = (i_1, \ldots, i_k), \quad n_j^{(k)} = (j_1, \ldots, j_k),$$

then

$$\chi_{ij} = (-1)^{\sum_{1 \leq r \leq k} i_r + \sum_{1 \leq s \leq k} j_s}.$$

BASIC Programs

1 Maximal replacement

```
'      MR.BAS
DATA   Maximal Replacement

'    linear equations, rank and base
'    inverse, determinant

DEFINT F-N: READ MRA$: Esc$ = CHR$(27): CR$ = CHR$(13)
READ d, t, o: DATA 1, 8, .00001: ' det, tab, round 0
CLS : PRINT MRA$; CR$: INPUT "Which problem? (1-3) "; i
ON i GOSUB Prob1, Prob2, Prob3

READ Prob$: CLS : PRINT MRA$; CR$: READ t$:
WHILE t$ <> "*": PRINT t$: READ t$: WEND: GOSUB Pause
READ m, n, k: l = m: IF m < n THEN l = n

fe = (k = 1)        ' equations
IF k = -1 THEN fi = -1: k = m        ' inverse
fh = (k = 0)        ' homogeneous

DIM a0(m, n + k), a(m, n + k)        ' orig & work arrays
DIM g(m), h(n)     ' replacement regs
DIM x(n), y(n, n)     ' particular & reduced sys solns
DIM r0(l), r1(m), c0(l), c1(n)        ' row, col base & non-base
DIM p(1, l, l)     ' critical submatrix

GOSUB Main: END
Main:
FOR i = 1 TO m: FOR j = 1 TO n - fe
  READ a(i, j): a0(i, j) = a(i, j)
NEXT j, i
IF fi THEN FOR i = 1 TO m: a(i, n + i) = 1: NEXT
GOSUB NewTable:        ' 1st table
GOSUB Find:        ' max replacement
GOSUB Results:        ' rank &c

IF fi THEN GOSUB SubInv
IF fe THEN GOSUB GenSys
IF fh THEN GOSUB RedSys

GOSUB Pause: RETURN

Find: '___ find replaceable generator
IF r = m THEN RETURN            ' none, end
FOR r = r + 1 TO m              ' search further
  IF g(r) = 0 THEN GOTO rFound  ' otherwise try next
```

```
NEXT: RETURN                              ' if none, end
rFound: '___ now find possible replacement
  FOR s = 1 TO n
    IF ABS(a(r, s)) > o THEN GOTO sFound      ' otherwise next
  NEXT: RETURN                              ' if none, end

sFound:
  d = d * a(r, s)          ' determinant
  GOSUB NewRep             ' new replacement

'__Replace
  x = 1 / a(r, s)
  FOR j = 1 TO n + k
    a(r, j) = a(r, j) * x
  NEXT j
  FOR i = 1 TO m: IF i = r THEN GOTO Rec
    x = a(i, s)
    FOR j = 1 TO n + k
      a(i, j) = a(i, j) - x * a(r, j)
    NEXT j
Rec:
  NEXT i
  g(r) = s: h(s) = r          ' register replacement
  GOSUB NewTable: GOSUB Pause    ' new table
  r = 0: GOTO Find             ' loop

'___Print subs

RowLbl:
  IF g(i) THEN i$ = "a": g = g(i) ELSE i$ = "e": g = i
  i$ = i$ + MID$(STR$(g), 2): PRINT i$; : RETURN

ColLbl:
  FOR j = 1 TO n
    j$ = "a" + MID$(STR$(j), 2)
    PRINT TAB(6 + j * t); j$;
  NEXT: IF k = 0 THEN PRINT : RETURN

AdjCol: '___ Adjunct col labels
  IF fi THEN l$ = "e" ELSE l$ = "q"
  FOR j = 1 TO k
    j$ = l$ + MID$(STR$(j), 2)
    PRINT TAB(6 + (j + n) * t); j$;
  NEXT: PRINT : RETURN

NewTable:
  GOSUB ColLbl
  FOR i = 1 TO m
    GOSUB RowLbl
    FOR j = 1 TO n + k
      a = a(i, j): IF ABS(a) < o THEN a = 0   ' round to 0
      a(i, j) = a: PRINT TAB(4 + j * t); a;
    NEXT: PRINT
  NEXT: PRINT : RETURN

NewRep: '___ new replacement
  r$ = MID$(STR$(r), 2): s$ = MID$(STR$(s), 2)
  PRINT "replace generator e"; r$; " by a"; s$; CR$: RETURN

Results:

'__Rank
```

```
q = 0
FOR i = 1 TO m
  q = q - (g(i) > 0)
NEXT: PRINT "rank A ="; q; CR$

'__Base
  PRINT "row base: "; CR$, : j0 = 0: j1 = 0
  FOR i = 1 TO m
    IF g(i) > 0 THEN j0 = j0 + 1: r0(j0) = i: PRINT i;
    IF g(i) = 0 THEN j1 = j1 + 1: r1(j1) = i
  NEXT
  PRINT : PRINT "column base: "; CR$, : j0 = 0: j1 = 0
  FOR i = 1 TO n
    IF h(i) > 0 THEN j0 = j0 + 1: c0(j0) = i: PRINT i;
    IF h(i) = 0 THEN j1 = j1 + 1: c1(j1) = i
  NEXT: PRINT CR$
  PRINT "determinant = "; d; CR$: RETURN

GenSys: '___ general system
  FOR i = 1 TO m - q
    IF a(r1(i), n + 1) THEN PRINT "no solution": RETURN
  NEXT: fx = -1      ' soln exists
  FOR i = 1 TO q: x(g(r0(i))) = a(r0(i), n + 1): NEXT
  PRINT "particular solution: "; CR$,
  FOR i = 1 TO n: PRINT x(i); : NEXT: PRINT

RedSys: '___ reduced system
  IF q = n THEN PRINT "solution unique": RETURN
  FOR i = 1 TO q: FOR j = 1 TO n - q
    y(g(r0(i)), j) = a(r0(i), c1(j))
  NEXT j, i
  FOR j = 1 TO n - q
    y(c1(j), j) = -1
  NEXT
  PRINT CR$; "reduced system, solution base: "
  FOR j = 1 TO n - q: PRINT j,
    FOR i = 1 TO n
      PRINT y(i, j);
    NEXT: PRINT
  NEXT: RETURN

SubInv: '___ critical submatrix & inverse
  FOR i = 1 TO q: FOR j = 1 TO q
    p(0, i, j) = a0(h(c0(i)), c0(j))
    p(1, i, j) = a(h(c0(i)), n + c0(j))
  NEXT j, i

MatInv:
  FR = (q = m AND q = n): f = 0
  IF NOT FR THEN PRINT "critical sub";
  PRINT "matrix: "
  FOR j = 1 TO q
    PRINT TAB(6 + j * t); c0(j);
  NEXT: PRINT
  FOR i = 1 TO q: PRINT h(c0(i));
    FOR j = 1 TO q
      PRINT TAB(4 + j * t); p(-f, i, j);
    NEXT: PRINT
  NEXT: PRINT CR$: f = NOT f
  PRINT "inverse:"
```

```
FOR j = 1 TO q
   PRINT TAB(6 + j * t); c0(j);
NEXT: PRINT
FOR i = 1 TO q: PRINT h(c0(i));
   FOR j = 1 TO q
      PRINT TAB(4 + j * t); a(r0(i), n + c0(j));
   NEXT: PRINT
NEXT: PRINT CR$: RETURN
```

Pause:
```
   k$ = INPUT$(1): IF k$ = Esc$ THEN END ELSE RETURN
```

```
Prob1: RESTORE Dat1: RETURN
Prob2: RESTORE Dat2: RETURN
Prob3: RESTORE Dat3: RETURN
```

```
Dat1: DATA PROB1, Problem 1
      DATA "Test example, inverse",,*
      DATA 3, 3, -1
      DATA 1, 2, 3
      DATA 3, 4, 5
      DATA 2, 3, 8
```

```
Dat2: DATA PROB2, Problem 2
      DATA "Test example, equations",unique solution,,*
      DATA 3, 3, 1
      DATA 1, 2, 3,  6
      DATA 3, 4, 5, 12
      DATA 2, 3, 8, 13
```

```
Dat3: DATA PROB3, Problem 3
      DATA "Test example, equations",many solutions,,*
      DATA 3, 3, 1
      DATA 1, 2, 3,  6
      DATA 3, 4, 7, 14
      DATA 2, 3, 5, 10
```

DEMONSTRATIONS:

Maximal Replacement

Problem 1
Test example, inverse

	a1	a2	a3	e1	e2	e3
e1	1	2	3	1	0	0
e2	3	4	5	0	1	0
e3	2	3	8	0	0	1

replace generator e1 by a1

	a1	a2	a3	e1	e2	e3
a1	1	2	3	1	0	0
e2	0	-2	-4	-3	1	0
e3	0	-1	2	-2	0	1

replace generator e2 by a2

	a1	a2	a3	e1	e2	e3
a1	1	0	-1	-2	1	0
a2	0	1	2	1.5	-.5	0
e3	0	0	4	-.5	-.5	1

replace generator e3 by a3

	a1	a2	a3	e1	e2	e3
a1	1	0	0	-2.125	.875	.25
a2	0	1	0	1.75	-.25	-.5
a3	0	0	1	-.125	-.125	.25

rank A = 3
row base: 1 2 3
column base: 1 2 3
determinant = -8

matrix:

	1	2	3
1	1	2	3
2	3	4	5
3	2	3	8

inverse

	1	2	3
1	-2.125	.875	.25
2	1.75	-.25	-.5
3	-.125	-.125	.25

Problem 2
Test example, equations

	a1	a2	a3	q1
e1	1	2	3	6
e2	3	4	5	12
e3	2	3	8	13

replace generator e1 by a1

	a1	a2	a3	q1
a1	1	2	3	6
e2	0	-2	-4	-6
e3	0	-1	2	1

replace generator e2 by a2

	a1	a2	a3	q1
a1	1	0	-1	0
a2	0	1	2	3
e3	0	0	4	4

replace generator e3 by a3

	a1	a2	a3	q1
a1	1	0	0	1
a2	0	1	0	1
a3	0	0	1	1

rank A = 3

row base: 1 2 3
column base: 1 2 3

determinant = -8

particular solution: 1 1 1
solution unique

Problem 3
Test example, equations
many solutions

	a1	a2	a3	q1
e1	1	2	3	6
e2	3	4	7	14
e3	2	3	5	10

replace generator e1 by a1

	a1	a2	a3	q1
a1	1	2	3	6
e2	0	-2	-2	-4
e3	0	-1	-1	-2

replace generator e2 by a2

	a1	a2	a3	q1
a1	1	0	1	2
a2	0	1	1	2
e3	0	0	0	0

rank A = 2
row base: 1 2
column base: 1 2

determinant = -2

particular solution:
    2 2 0

reduced system, solution base:
1   1 1 -1

2 Rank reduction

```
'     RR.BAS
DATA   Rank Reduction

'     matrix rank, row & col bases, determinant
   DEFINT F-N, T-V: Esc$ = CHR$(27): CR$ = CHR$(13)
   CLS : READ rra$: PRINT rra$; CR$
   READ t, t0, d, o: DATA 12, 4, 1, .000001

   READ m, n: DIM a(m, n), b(m, n), i(m), j(n)
   FOR i = 1 TO m: FOR j = 1 TO n
     READ b(i, j)
   NEXT j, i

   GOSUB bTOa: GOSUB Results: GOSUB Pause: END

Find: '__ find reduction element
   FOR i = 1 TO m
     FOR j = 1 TO n
       IF a(i, j) THEN GOTO Found
     NEXT
   NEXT: RETURN

Found:
   h = h + 1: i(h) = i: j(h) = j: p = a(i, j)
   u = i: v = j: r = 1 / p: d = d * p
   PRINT "element "; p; " in row "; u; " column "; v; CR$

'__reduce a to b
   FOR i = 1 TO m: FOR j = 1 TO n
     b(i, j) = a(i, j) - a(i, v) * r * a(u, j)
   NEXT j, i
```

```
bTOa: '___ copy b to a & loop
   FOR i = 1 TO m: FOR j = 1 TO n
     b = b(i, j): IF ABS(b) <= o THEN b = 0
     a(i, j) = b: PRINT TAB(j * t - t0); b;
   NEXT: PRINT : NEXT: PRINT : GOSUB Pause: GOTO Find
```

```
Results:
   PRINT "rank ="; h
   PRINT "row base:",
     FOR k = 1 TO h
       PRINT i(k);
     NEXT: PRINT
   PRINT "column base:",
     FOR k = 1 TO h
       PRINT j(k);
     NEXT: PRINT
   PRINT "determinant = "; d: RETURN
```

```
Pause:
   k$ = INPUT$(1): IF k$ = Esc$ THEN END ELSE RETURN
```

```
'__Problem
   DATA 4, 3          :' m, n
     DATA 4, 5, 1    :' b11, ...
     DATA 2, -3, -5  :' ...
     DATA 2, -4, -6
     DATA -1, 2, 3
```

DEMONSTRATION:

Rank Reduction

```
4      5      1
2     -3     -5
2     -4     -6
-1      2      3
```

element 4 in row 1 column 1

```
0      0      0
0     -5.5   -5.5
0     -6.5   -6.5
0      3.25   3.25
```

element -5.5 in row 2 column 2

```
0      0      0
0      0      0
0      0      0
0      0      0
```

rank = 2
row base: 1 2
column base: 1 2
determinant = -22

3 Tucker's Pivot Algorithm

```
'     TPA.BAS
DATA   Tucker Pivot Algorithm
```

```
'     linear equations, rank & base, inverse, determinant
```

```
    DEFINT F-N: Esc$ = CHR$(27): CR$ = CHR$(13)
    READ tpa$: CLS : PRINT tpa$; CR$
    READ d, t, o: DATA 1, 8, .000001

    INPUT "which problem? (1-3) "; i
    ON i GOSUB Prob1, Prob2, Prob3
    READ Prob$: CLS : PRINT tpa$; CR$
    READ t$: WHILE t$ <> "*": PRINT t$: READ t$: WEND
    READ m, n, m0, n0
    fe = (m0 = 1 AND n0 = 0)    ' equations

    DIM a0(m, n), a(m, n), b(m, n), g(m), h(n)
    FOR i = m0 TO m: FOR j = n0 TO n
        READ a0(i, j): b(i, j) = a0(i, j)
    NEXT j, i
    FOR j = 1 TO n: h(j) = j: NEXT        'basic are 1 to N
    FOR i = 1 TO m: g(i) = n + i: NEXT    'non-basic are N+1 to N+M


    GOSUB BtoA        ' max pivot
    GOSUB Results     ' rank &c
    IF fe THEN GOSUB Eqns
    GOSUB Pause: END

Find: '___ find pivot element
    FOR r = 1 TO m: FOR s = 1 TO n
        IF g(r) > n AND h(s) <= n AND a(r, s) THEN GOTO Found
    NEXT s, r: RETURN

Found:
    p = a(r, s): d = d * p      ' determinant
    PRINT "pivot element "; p; "in row"; r; "column"; s; CR$

'___pivot on A produces new tableau B
    FOR i = m0 TO m: FOR j = n0 TO n
        b(i, j) = a(i, j) - a(i, s) * a(r, j) / p
        IF i = r THEN b(i, j) = a(i, j) / p
        IF j = s THEN b(i, j) = -a(i, j) / p
        IF i = r AND j = s THEN b(i, j) = 1 / p
    NEXT j, i: SWAP g(r), h(s)    ' register exchange

BtoA: '___ B becomes A for another loop
    GOSUB BasLabel: IF m0 THEN fb = 0
    FOR i = m0 TO m: GOSUB NonBasLabel
        FOR j = n0 TO n
        a = b(i, j): IF ABS(a) < o THEN a = 0
        a(i, j) = a: PRINT TAB(4 + j * t); a;
        NEXT: PRINT
    NEXT: PRINT : GOSUB Pause: GOTO Find

Variable: '___ identify variables, basic or non-basic
    IF fb THEN v = h(j) ELSE v = g(i)
    IF v > n THEN v$ = "Y": v = v - n ELSE v$ = "X"
    v$ = v$ + MID$(STR$(v), 2): RETURN

BasLabel: '___ basic labels
    fb = 1
    FOR j = n0 TO n: GOSUB Variable
        PRINT TAB(6 + j * t); v$;
    NEXT: PRINT : RETURN

NonBasLabel: '___ non-basic labels
    IF i THEN GOSUB Variable: PRINT v$; ELSE fb = 0
```

```
    RETURN
Results: '___ rank &c
    FOR i = 1 TO m
      q = q - (g(i) <= n)
    NEXT: PRINT "rank ="; q
    DIM r0(q), r1(m - q), s0(q), s1(n - q), p(q, q), pi(q, q)
    FR = (q = m AND q = n)
    I0 = 0: I1 = 0
    FOR i = 1 TO m
      IF g(i) <= n THEN I0 = I0 + 1: r0(I0) = i
      IF g(i) > n THEN I1 = I1 + 1: r1(I1) = i
    NEXT
    I0 = 0: I1 = 0
    FOR i = 1 TO n
      IF h(i) > n THEN I0 = I0 + 1: s0(I0) = i
      IF h(i) <= n THEN I1 = I1 + 1: s1(I1) = i
    NEXT

RCBases:
    PRINT "row base:",
      FOR i = 1 TO q
        PRINT g(r0(i));
      NEXT: PRINT
    PRINT "column base:",
      FOR i = 1 TO q
        PRINT h(r0(i)) - n;
      NEXT: PRINT
    IF f0 THEN RETURN

    IF FR THEN m$ = "matrix" ELSE m$ = "pivot block"
      PRINT m$
      FOR i = 1 TO q
        FOR j = 1 TO q
          p(i, j) = a0(r0(i), s0(j))
          PRINT TAB(4 + j * t); p(i, j);
        NEXT: PRINT
      NEXT

    PRINT "inverse"
      FOR i = 1 TO q
        FOR j = 1 TO q
          pi(i, j) = a(r0(i), s0(j))
          PRINT TAB(4 + j * t); pi(i, j);
        NEXT: PRINT
      NEXT: PRINT

    PRINT "determinant = "; d; CR$

'__Sort bases
    FOR i = 1 TO q - 1 STEP -1: f = 0
      FOR j = 1 TO i
IF g(r0(j)) > g(r0(j + 1)) THEN SWAP r0(j), r0(j + 1): f = -1
      NEXT: IF f = 0 THEN GOTO RDone ELSE f0 = -1
    NEXT
RDone:
    FOR i = 1 TO q - 1 STEP -1: f = 0
      FOR j = 1 TO i
IF h(s0(j)) > h(s0(j + 1)) THEN SWAP s0(j), s0(j + 1): f = -1
      NEXT: IF f = 0 THEN GOTO CDone ELSE f0 = -1
    NEXT
```

```
CDone:
  IF f0 THEN PRINT "sorted bases": GOSUB RCBases
  RETURN

Eqns: '___ equations
  FOR i = 1 TO m - q
    IF a(g(r1(i)) - n, 0) THEN PRINT "no solution": RETURN
  NEXT: DIM x(n)

  PRINT "particular solution:"; CR$: fb = 0
    FOR r = 1 TO q: i = g(r0(r)): GOSUB Variable
      PRINT , v$; " ="; a(i, 0): x(r0(r)) = a(i, 0)
    NEXT: fb = 1
    FOR s = 1 TO n - q: j = h(s1(s)): GOSUB Variable
      PRINT , v$; " ="; 0
    NEXT: PRINT

  PRINT "column nullity ="; n - q
    IF q = n THEN PRINT "unique solution": RETURN
    PRINT "reduced system, solution base:"; CR$
    DIM y(n, n - q)

    FOR i = 1 TO q: FOR j = 1 TO n - q
      y(r0(i), j) = a(g(r0(i)), h(s1(j)))
    NEXT j, i
    FOR j = 1 TO n - q
      y(h(s1(j)), j) = -1
    NEXT j

    FOR j = 1 TO n - q: PRINT j; ")";
      FOR i = 1 TO n
        PRINT TAB(4 + i * t); y(i, j);
      NEXT: PRINT
    NEXT
  RETURN

Pause:  k$ = INPUT$(1): IF k$ = Esc$ THEN END ELSE RETURN

Prob1: RESTORE Dat1: RETURN
Prob2: RESTORE Dat2: RETURN
Prob3: RESTORE Dat3: RETURN

Dat1: DATA PROB1, Problem 1
  DATA "Test example, inverse",,*
  DATA 3, 3, 1, 1
    DATA 1, 2, 3
    DATA 3, 4, 5
    DATA 2, 3, 8

Dat2: DATA PROB2, Problem 2
  DATA "Test example, equations",unique solution,,*
  DATA 3, 3, 1, 0
    DATA 6, 1, 2, 3
    DATA 12, 3, 4, 5
    DATA 13, 2, 3, 8

Dat3: DATA PROB3, Problem 3
  DATA "Test example, equations",many solutions,,*
  DATA 3, 3, 1, 0
    DATA 6, 1, 2, 3
    DATA 14, 3, 4, 7
    DATA 10, 2, 3, 5
```

148

BASIC PROGRAMS

DEMONSTRATIONS:

Tucker's Algorithm

Problem 1
Test example, inverse

	X1	X2	X3
Y1	1	2	3
Y2	3	4	5
Y3	2	3	8

pivot element 1 in row 1 column 1

	Y1	X2	X3
X1	1	2	3
Y2	-3	-2	-4
Y3	-2	-1	2

pivot element -2 in row 2 column 2

	Y1	Y2	X3
X1	-2	1	-1
X2	1.5	-.5	2
Y3	-.5	-.5	4

pivot element 4 in row 3 column 3

	Y1	Y2	Y3
X1	-2.125	.875	.25
X2	1.75	-.25	-.5
X3	-.125	-.125	.25

rank = 3
row base: 1 2 3
column base: 1 2 3

matrix

```
1   2   3
3   4   5
2   3   8
```

inverse
```
-2.125  .875  .25
 1.75  -.25  -.5
 -.125 -.125  .25
```

determinant = -8

Problem 2
Test example, equations
unique solution

	X0	X1	X2	X3
Y1	6	1	2	3
Y2	12	3	4	5
Y3	13	2	3	8

pivot element 1 in row 1 column 1

	X0	Y1	X2	X3
X1	6	1	2	3
Y2	-6	-3	-2	-4
Y3	1	-2	-1	2

pivot element -2 in row 2 column 2

	X0	Y1	Y2	X3
X1	0	-2	1	-1
X2	3	1.5	-.5	2
Y3	4	-.5	-.5	4

pivot element 4 in row 3 column 3

	X0	Y1	Y2	Y3
X1	1	-2.125	.875	.25
X2	1	1.75	-.25	-.5
X3	1	-.125	-.125	.25

rank = 3
row base: 1 2 3
column base: 1 2 3

matrix

1	2	3
3	4	5
2	3	8

inverse

-2.125	.875	.25
1.75	-.25	-.5
-.125	-.125	.25

determinant = -8

particular solution:
$$X1 = 1$$
$$X2 = 1$$
$$X3 = 1$$

column nullity = 0
unique solution

Problem 3
Test example, equations
many solutions

	X0	X1	X2	X3
Y1	6	1	2	3
Y2	14	3	4	7
Y3	10	2	3	5

pivot element 1 in row 1 column 1

	X0	Y1	X2	X3
X1	6	1	2	3
Y2	-4	-3	-2	-2
Y3	-2	-2	-1	-1

pivot element -2 in row 2 column 2

	X0	Y1	Y2	X3
X1	2	-2	1	1
X2	2	1.5	-.5	1
Y3	0	-.5	-.5	0

rank = 2
row base: 1 2
column base: 1 2

pivot block

```
      1    2
      3    4
```
inverse
```
     -2    1
     1.5  -.5
```
determinant = -2

particular solution:
```
        X1 = 2
        X2 = 2
        X3 = 0
```
column nullity = 1

reduced system, solution base:
```
  1 )    1    1    -1
```

4 Extended rank reduction

```
'    RRX.BAS
DATA    Extended Rank Reduction
'    rank & base, determinant, equations, inverse
    DEFINT F-N, T: ESC$ = CHR$(27): CR$ = CHR$(13)
    READ rax$: CLS : PRINT rax$; CR$
    READ d, t, O: DATA 1, 8, .000001
    INPUT "which problem? (0-3) "; i
    ON i + 1 GOSUB Prob0, Prob1, Prob2, Prob3: READ Prob$

    CLS : PRINT rax$; CR$: READ t$
    WHILE t$ <> "*": PRINT t$: READ t$: WEND
    READ m, n, n0, fi

    k = fi * m          ' inverse
    fe = (n0 = 0)       ' equations
    DIM a0(m, n), a(m, n + k), b(m, n + k), c(n, n + k)
    DIM h(m), k(n), u(n, n), v(n, n)

    FOR i = 1 TO n: v(i, i) = 1: NEXT
    FOR i = 1 TO m: FOR j = n0 TO n
        READ a0(i, j): b(i, j) = a0(i, j)
    NEXT j, i
    IF fi THEN FOR i = 1 TO m: b(i, n + i) = 1: NEXT

    GOSUB BtoA          ' reduce
    GOSUB Results       ' rank &c
    IF fi THEN GOSUB Inv
    IF fe THEN GOSUB Eqns
    GOSUB Pause: END
PrintC:
  PRINT TAB(8); "C"
  FOR i = 1 TO n: FOR j = n0 TO n + k
      PRINT TAB(4 + (j + 1 - n0) * t); c(i, j);
  NEXT j: PRINT : NEXT: PRINT : RETURN

PrintU:
  PRINT TAB(8); "U"
  FOR i = 1 TO n: FOR j = 1 TO n
      PRINT TAB(4 + j * t); u(i, j);
```

```
NEXT j: PRINT : NEXT: PRINT : RETURN

Find: '___ Find reduction element
    FOR r = 1 TO m: FOR s = 1 TO n
        IF a(r, s) THEN GOTO Found
    NEXT s, r: RETURN

Found:
    q = q + 1: h(r) = 1: k(s) = 1: p = a(r, s)
    PRINT "element "; p; " in row "; r; " column "; s; CR$
    d = d * p: p = 1 / p: '___ determinant D
'__reduce A to B
    FOR j = n0 TO n + k: c = p * a(r, j)
    FOR i = 1 TO m
    b(i, j) = a(i, j) - a(i, s) * c: NEXT i, j
'__compensate C
    FOR j = n0 TO n + k: c = p * a(r, j)
    FOR i = 1 TO n: c(i, j) = c(i, j) + u(i, s) * c
    NEXT i, j
'__update U to V
    FOR j = 1 TO n: c = p * a(r, j)
    FOR i = 1 TO n: v(i, j) = u(i, j) - u(i, s) * c
    NEXT i, j
BtoA: '___ copy B to A
    FOR i = 1 TO m: FOR j = n0 TO n + k
        b = b(i, j): IF ABS(b) <= O THEN b = 0: ' 0-error
        a(i, j) = b: PRINT TAB(4 + (j + 1 - n0) * t); b;
    NEXT: PRINT : NEXT: PRINT
'__copy V to U
    FOR i = 1 TO n: FOR j = 1 TO n
    u(i, j) = v(i, j): NEXT j, i
'__test U & C
    GOSUB PrintC: GOSUB PrintU
'__loop
    GOSUB Pause
    GOTO Find

Results: '___ rank &c
    PRINT "rank ="; q
    DIM r0(q), r1(m - q), s0(q), s1(n - q), p(q, q), pi(q, q)
    FR = (q = m AND q = n)

    FOR i = 1 TO m
        IF h(i) = 1 THEN i0 = i0 + 1: r0(i0) = i
        IF h(i) = 0 THEN i1 = i1 + 1: r1(i1) = i
    NEXT
    FOR j = 1 TO n
        IF k(j) = 1 THEN j0 = j0 + 1: s0(j0) = j
        IF k(j) = 0 THEN j1 = j1 + 1: s1(j1) = j
    NEXT

    PRINT "row base:",
        FOR i = 1 TO q: PRINT r0(i); : NEXT: PRINT
    PRINT "column base:",
        FOR j = 1 TO q: PRINT s0(j); : NEXT: PRINT

    IF FR THEN m$ = "matrix" ELSE m$ = "reduction block"
        PRINT m$
        FOR i = 1 TO q: FOR j = 1 TO q
            a = a0(r0(i), s0(j)): p(i, j) = a
            PRINT TAB(4 + j * t); a;
```

```
    NEXT: PRINT : NEXT

    PRINT "determinant = "; d; CR$: RETURN
Inv:
    PRINT "inverse"
      FOR i = 1 TO q: FOR j = 1 TO q
          c = c(s0(i), n + s0(j)): pi(i, j) = c
          PRINT TAB(4 + j * t); c;
      NEXT: PRINT : NEXT: PRINT : RETURN

Eqns:
      FOR i = 1 TO m - q
          IF a(r1(i), 0) THEN PRINT "no solution": RETURN
      NEXT: DIM x(n)
      PRINT "particular solution:"; CR$,
          FOR i = 1 TO q: x(r0(i)) = c(r0(i), 0): NEXT
          FOR i = 1 TO n: PRINT x(i); : NEXT: PRINT

      PRINT "column nullity ="; n - q
      IF q = n THEN PRINT "unique solution": RETURN
      PRINT "reduced system, solution base:"; CR$
      DIM y(n, n - q)

      FOR i = 1 TO q: FOR j = 1 TO n - q
          y(r0(i), j) = c(r0(i), s1(j))
      NEXT j, i
      FOR j = 1 TO n - q: y(s1(j), j) = -1: NEXT j

      FOR j = 1 TO n - q: PRINT j; ")"; : FOR i = 1 TO n
          PRINT TAB(4 + i * t); y(i, j);
      NEXT: PRINT : NEXT

    RETURN

Pause: k$ = INPUT$(1): IF k$ = ESC$ THEN END ELSE RETURN

Prob0: RESTORE Dat0: RETURN
Prob1: RESTORE Dat1: RETURN
Prob2: RESTORE Dat2: RETURN
Prob3: RESTORE Dat3: RETURN

Dat0: DATA PROB0, Problem 0
      DATA Test example, rank &c,,*
      DATA 4, 3, 1, 0    : '  m, n, n0, fi
       DATA 4, 5, 1 : '  b11, ...
       DATA 2, -3, -5 : '  ...
       DATA 2, -4, -6
       DATA -1, 2, 3

Dat1: DATA PROB1, Problem 1
      DATA "Test example, inverse",,*
       DATA 3, 3, 1, 1
       DATA 1, 2, 3
       DATA 3, 4, 5
       DATA 2, 3, 8

Dat2: DATA PROB2, Problem 2
      DATA "Test example, equations",unique solution,,*
       DATA 3, 3, 0, 0
       DATA 6, 1, 2, 3
       DATA 12, 3, 4, 5
       DATA 13, 2, 3, 8
```

Dat3: DATA PROB3, Problem 3
    DATA "Test example, equations",many solutions,,*
    DATA 3, 3, 0, 0
      DATA 6, 1, 2, 3
      DATA 14, 3, 4, 7
      DATA 10, 2, 3, 5

DEMONSTRATIONS:

Rank Reduction Algorithm--extended

Problem 0
Test example
rank &c

```
        4     5     1
        2    -3    -5
        2    -4    -6
       -1     2     3
    C
        0     0     0
        0     0     0
        0     0     0
    U
        1     0     0
        0     1     0
        0     0     1
```

element 4 in row 1 column 1

```
        0     0     0
        0   -5.5  -5.5
        0   -6.5  -6.5
        0    3.25  3.25
    C
        1    1.25   .25
        0     0     0
        0     0     0
    U
        0   -1.25  -.25
        0     1     0
        0     0     1
```

element -5.5 in row 2 column 2

```
        0     0     0
        0     0     0
        0     0     0
        0     0     0
    C
        1     0    -1
        0     1     1
        0     0     0
    U
        0     0     1
        0     0    -1
        0     0     1
```

rank = 2
row base: 1 2
column base: 1 2

reduction block
```
     4    5
     2   -3
```

determinant = -22

Problem 1
Test example, inverse
```
     1    2    3    1    0    0
     3    4    5    0    1    0
     2    3    8    0    0    1
C
     0    0    0    0    0    0
     0    0    0    0    0    0
     0    0    0    0    0    0
U
          1    0    0
          0    1    0
          0    0    1
```

element 1 in row 1 column 1
```
     0    0    0    0    0    0
     0   -2   -4   -3    1    0
     0   -1    2   -2    0    1
C
     1    2    3    1    0    0
     0    0    0    0    0    0
     0    0    0    0    0    0
U
          0   -2   -3
          0    1    0
          0    0    1
```

element -2 in row 2 column 2
```
     0    0    0    0    0    0
     0    0    0    0    0    0
     0    0    4   -.5   -.5   1
C
     1    0   -1   -2    1    0
     0    1    2   1.5   -.5   0
     0    0    0    0    0    0
U
          0    0    1
          0    0   -2
          0    0    1
```

element 4 in row 3 column 3
```
     0    0    0    0    0    0
     0    0    0    0    0    0
     0    0    0    0    0    0
C
     1    0    0  -2.125  .875   .25
     0    1    0    1.75  -.25   -.5
     0    0    1   -.125 -.125   .25
U
          0    0    0
          0    0    0
          0    0    0
```

rank = 3
row base: 1 2 3
column base: 1 2 3

matrix

1	2	3
3	4	5
2	3	8

determinant = -8

inverse

-2.125	.875	.25
1.75	-.25	-.5
-.125	-.125	.25

Problem 2
Test example, equations
unique solution

6	1	2	3
12	3	4	5
13	2	3	8

C

0	0	0	0
0	0	0	0
0	0	0	0

U

1	0	0
0	1	0
0	0	1

element 1 in row 1 column 1

0	0	0	0
-6	0	-2	-4
1	0	-1	2

C

6	1	2	3
0	0	0	0
0	0	0	0

U

0	-2	-3
0	1	0
0	0	1

element -2 in row 2 column 2

0	0	0	0
0	0	0	0
4	0	0	4

C

0	1	0	-1
3	0	1	2
0	0	0	0

U

0	0	1
0	0	-2
0	0	1

element 4 in row 3 column 3

0	0	0	0

```
        0    0    0    0
        0    0    0    0
C
        1    1    0    0
        1    0    1    0
        1    0    0    1
U
        0    0    0
        0    0    0
        0    0    0
```

rank = 3
row base: 1 2 3
column base: 1 2 3

matrix
```
        1    2    3
        3    4    5
        2    3    8
```

determinant = -8

particular solution:
       1   1   1

column nullity = 0
unique solution

Problem 3
Test example, equations
many solutions

```
        6    1    2    3
       14    3    4    7
       10    2    3    5
C
        0    0    0    0
        0    0    0    0
        0    0    0    0
U
        1    0    0
        0    1    0
        0    0    1
```

element 1 in row 1 column 1

```
        0    0    0    0
       -4    0   -2   -2
       -2    0   -1   -1
C
        6    1    2    3
        0    0    0    0
        0    0    0    0
U
        0   -2   -3
        0    1    0
        0    0    1
```

element -2 in row 2 column 2

```
        0    0    0    0
        0    0    0    0
        0    0    0    0
```

```
C
   2    1    0    1
   2    0    1    1
   0    0    0    0
U
   0    0   -1
   0    0   -1
   0    0    1
```

rank = 2
row base: 1 2
column base: 1 2

reduction block
        1   2
        3   4

determinant = -2

particular solution:
         2 2 0

column nullity = 1

reduced system, solution base:
  1)   1   1   -1

5 Permutations

5.1 Permutations and inverses

```
'     PERMI.BAS
'     Listing the Permutations
'     and their inverses
'     (transposition method)

DEFINT A-Z: CLS : INPUT "n ="; n
DIM L(n), M(n): t = n: f = 1
FOR i = 1 TO n
   L(i) = i: M(i) = i: f = f * i
NEXT
PRINT "The"; f; "permutations on"; n; "objects, and their inverses"
PRINT : GOSUB Pr

0 :
   IF L(t) > 1 THEN GOSUB 1: GOTO 0
   FOR L = 1 TO t - 1
      M(L) = M(L + 1): L(M(L)) = L
   NEXT: M(t) = t: L(t) = t: t = t - 1
   IF t > 1 THEN GOTO 0 ELSE END

1 :
   L = L(t): M = M(L - 1)
   L(M) = L: M(L) = M: L(t) = L - 1: M(L - 1) = t: t = n
Pr: q = q + 1: PRINT q; TAB(8);
   FOR i = 1 TO n: PRINT M(i); : NEXT: PRINT SPC(3);
   FOR i = 1 TO n: PRINT L(i); : NEXT: PRINT : RETURN

n =? 4
The 24 permutations on 4 objects, and their inverses

 1    1 2 3 4   1 2 3 4
```

2	1 2 4 3	1 2 4 3
3	1 4 2 3	1 3 4 2
4	4 1 2 3	2 3 4 1
5	1 3 2 4	1 3 2 4
6	1 3 4 2	1 4 2 3
7	1 4 3 2	1 4 3 2
8	4 1 3 2	2 4 3 1
9	3 1 2 4	2 3 1 4
10	3 1 4 2	2 4 1 3
11	3 4 1 2	3 4 1 2
12	4 3 1 2	3 4 2 1
13	2 1 3 4	2 1 3 4
14	2 1 4 3	2 1 4 3
15	2 4 1 3	3 1 4 2
16	4 2 1 3	3 2 4 1
17	2 3 1 4	3 1 2 4
18	2 3 4 1	4 1 2 3
19	2 4 3 1	4 1 3 2
20	4 2 3 1	4 2 3 1
21	3 2 1 4	3 2 1 4
22	3 2 4 1	4 2 1 3
23	3 4 2 1	4 3 1 2
24 ,	4 3 2 1	4 3 2 1

5.2 Permutations in order of last differences

```
'     PERMR.BAS
'     Listing the Permutations
'     order of last differences
'     (recursion)
DEFINT A-Z: CLS : INPUT "n ="; n: DIM p(n, n), s(n): nn = n
f = 1: FOR i = 1 TO n: f = f * i: p(n, i) = i: NEXT
PRINT "The"; f; "permutations on"; n; "objects"
PRINT "in order of last differing elements": PRINT

      GOSUB O: END

O:
   IF n = 1 THEN GOSUB Pr: RETURN
   FOR m = n TO 1 STEP -1: GOSUB X
      s(n) = m: n = n - 1
      GOSUB O
      n = n + 1: m = s(n)
   NEXT: RETURN

X :
   SWAP p(n, m), p(n, n)
   FOR i = 1 TO nn
      p(n - 1, i) = p(n, i)
   NEXT: RETURN

Pr:
   q = q + 1: PRINT q,
   FOR i = 1 TO nn
      PRINT p(1, i);
   NEXT: PRINT : RETURN


n =? 4
The 24 permutations on 4 objects
```

in order of last differing elements

1	1	2	3	4
2	2	1	3	4
3	1	3	2	4
4	3	1	2	4
5	2	3	1	4
6	3	2	1	4
7	1	2	4	3
8	2	1	4	3
9	1	4	2	3
10	4	1	2	3
11	2	4	1	3
12	4	2	1	3
13	1	3	4	2
14	3	1	4	2
15	1	4	3	2
16	4	1	3	2
17	3	4	1	2
18	4	3	1	2
19	2	3	4	1
20	3	2	4	1
21	2	4	3	1
22	4	2	3	1
23	3	4	2	1
24	4	3	2	1

5.3 Factorization into disjoint cycles

```
'    PCYC.BAS
'    Factorization of permutation
'    into disjoint cycles,
'    and alternating character

DEFINT A-Z: CLS : READ n: DIM a(n), b(n, n), p(n)
FOR i = 1 TO n: READ p(i): NEXT
'DATA n,   1, 2, 3, 4, 5, 6, 7, 8, 9, ...
 DATA 9,   2, 3, 1, 5, 6, 4, 7, 9, 8

O:
   GOSUB Another: IF k = 0 THEN GOSUB Pr: END
   i = k: j = 1: h = h + 1: b(h, j) = i: GOSUB Cycle
   GOTO O

Cycle:
   i = p(i): IF i = k THEN b(h, 0) = j: RETURN
   a(i) = 1: j = j + 1: b(h, j) = i: GOTO Cycle

Another:
   FOR k = 1 TO n
     IF a(k) = 0 THEN a(k) = 1: RETURN
   NEXT: k = 0: m = h: RETURN

Pr:
   PRINT "Permutation"
   FOR i = 1 TO n: PRINT p(i); : NEXT: PRINT
   PRINT "Cyclic factors"
   FOR h = 1 TO m: FOR j = 1 TO b(h, 0)
     PRINT b(h, j);
   NEXT j: PRINT : t = t + b(h, 0) - 1: NEXT h
   PRINT "Alternating character "; 1 - 2 * (t MOD 2)
```

RETURN

Permutation
 2 3 1 5 6 4 7 9 8
Cyclic factors
 1 2 3
 4 5 6
 7
 8 9
Alternating character -1

6 Combinations

6.1 Pascal's Triangle

```
'    PASCAL.BAS
DATA   Pascal's Triangle
'    (recursion)
DEFINT A-Z: READ f$: t = 5: cr$ = CHR$(13)
CLS : PRINT cr$; f$; cr$
INPUT "n = "; n: IF n = 0 THEN n = 12
n = n + 1: DIM p(n, n): p(0, 0) = 1: PRINT

    GOSUB Tri

k$ = INPUT$(1): IF k$ = cr$ THEN RUN ELSE END

Tri:
'  __triangle(n)
  IF n = 0 THEN RETURN
  n = n - 1: GOSUB Tri: n = n + 1
'  __line(n)
  FOR m = 1 TO n
    p(n, m) = p(n - 1, m - 1) + p(n - 1, m)
    PRINT TAB(m * t - t); p(n, m);
  NEXT: RETURN
```

```
Pascal's Triangle
n= 12

1
1   1
1   2   1
1   3   3   1
1   4   6   4   1
1   5   10  10  5   1
1   6   15  20  15  6   1
1   7   21  35  35  21  7   1
1   8   28  56  70  56  28  8   1
1   9   36  84  126 126 84  36  9   1
1   10  45  120 210 252 210 120 45  10  1
1   11  55  165 330 462 462 330 165 55  11  1
1   12  66  220 495 792 924 792 495 220 66  12  1
```

6.2 Binomial coefficients

```
'    B1.BAS
```

```
DATA   Binomial Coefficient
DATA    n objects m at a time
DEFINT A-Z: DEFLNG F
READ u$, v$: CLS : PRINT u$: PRINT v$: PRINT
I:
  INPUT "n = "; n: IF n = 0 THEN END
  INPUT "m = "; m: r = n - m: GOSUB O
  PRINT "binomial ("; n; ","; m; ") = "; f
  PRINT : CLEAR : GOTO I
O:
  IF r = 0 THEN f = 1: RETURN
  r = r - 1: GOSUB O: r = r + 1
  f = f * (m + r) \ r: RETURN


'    B2.BAS
DATA   Binomial Coefficient
DATA    n objects m at a time
DEFINT A-Z: READ u$, v$: CLS : PRINT u$: PRINT v$: PRINT
I:
  INPUT "n = "; n: IF n = 0 THEN END
  INPUT "m = "; m: DIM b(n, m) AS LONG: GOSUB O
  PRINT "binomial ("; n; ","; m; ") = "; b(n, m)
  PRINT : CLEAR : GOTO I
O:
  IF m = 0 OR m = n THEN b(n, m) = 1: RETURN
  n = n - 1: GOSUB O : m = m - 1: GOSUB O
  m = m + 1: n = n + 1
  b(n, m) = b(n - 1, m) + b(n - 1, m - 1) : RETURN

Binomial Coefficient
n objects m at a time

n = ? 12
m = ? 6
binomial ( 12 , 6 ) = 924

n = ? 31
m = ? 11
binomial ( 31 , 11 ) = 84672315
```

6.3 Combinations in order of last differences

```
'    C.BAS
DATA    Combinations of n objects m at a time
'         order of last differences
  DEFINT A-Z: READ t$: CLS : PRINT t$
  INPUT "n = "; n: IF n = 0 THEN END
  INPUT "m = "; m: DIM b(n, m) AS LONG
  GOSUB D: b = b(n, m)
  PRINT "binomial ("; n; ","; m; ") = "; b
  DIM c(n, m, b, m): GOSUB O: GOSUB Pr: END
O:
  IF m = 0 OR m = n THEN GOSUB Oo: RETURN
  n = n - 1: GOSUB O
  m = m - 1: GOSUB O
```

```
m = m + 1: n = n + 1
b0 = b(n - 1, m): b1 = b(n - 1, m - 1)
FOR i = 1 TO b0: FOR j = 1 TO m
   c(n, m, i, j) = c(n - 1, m, i, j)
NEXT j, i
FOR i = 1 TO b1: FOR j = 1 TO m - 1
   c(n, m, b0 + i, j) = c(n - 1, m - 1, i, j)
NEXT j: c(n, m, b0 + i, m) = n: NEXT i
RETURN

Oo:
   FOR j = 1 TO m: c(n, m, 1, j) = j: NEXT: RETURN

Pr:
   PRINT
   FOR i = 1 TO b: PRINT i,
      FOR j = 1 TO m
         PRINT c(n, m, i, j);
      NEXT j: PRINT
   NEXT i: PRINT : k$ = INPUT$(1): RETURN

D:
   IF m = 0 OR m = n THEN b(n, m) = 1: RETURN
   n = n - 1: GOSUB D
   m = m - 1: GOSUB D
   m = m + 1: n = n + 1
   b(n, m) = b(n - 1, m) + b(n - 1, m - 1)
   RETURN
```

Combinations of n objects m at a time
n = ? 6
m = ? 4
binomial (6 , 4) = 15

```
1         1 2 3 4
2         1 2 3 5
3         1 2 4 5
4         1 3 4 5
5         2 3 4 5
6         1 2 3 6
7         1 2 4 6
8         1 3 4 6
9         2 3 4 6
10         1 2 5 6
11         1 3 5 6
12         2 3 5 6
13         1 4 5 6
14         2 4 5 6
15         3 4 5 6
```

6.4 Location of a given combination

```
'    INDEX.BAS
DATA   "Combinations of 1, 2, ... , taken m at a time
DATA   indexed in order of last differences

   DEFINT A-Z: READ a$, b$: CLS : PRINT a$: PRINT b$: PRINT
In:
   INPUT "m = "; m: IF m THEN DIM c(m) ELSE END
   PRINT "enter 0 < c1 < ... < c" + MID$(STR$(m), 2): h = m
```

```
    FOR i = 1 TO m
Again:
    PRINT "c" + MID$(STR$(i), 2) + " = "; : INPUT c(i)
    IF c(i) <= c(i - 1) THEN PRINT "do again: "; : GOTO Again
    NEXT: GOSUB Index: CLEAR : GOTO In

Index:
    i = 1
    FOR m = m TO 1 STEP -1
    IF c(m) = m THEN GOSUB Fin: RETURN
    f = 1
    FOR r = 1 TO c(m) - m - 1
        f = f * (m + r) \ r
    NEXT: i = i + f
    NEXT: GOSUB Fin: RETURN

Fin:
    PRINT "The combination"
    FOR j = 1 TO h: PRINT c(j); : NEXT: PRINT
    PRINT "has index "; i: PRINT : RETURN
```

Combinations of 1, 2, ... , taken m at a time
indexed in order of last differences

```
m = ? 4
enter 0 < c1 < ... < c4
c1 = ? 1
c2 = ? 3
c3 = ? 5
c4 = ? 6
```

The combination
 1 3 5 6
has index 11

6.5 Combination in a given location

```
'      INV.BAS
DATA   Combinations m at a time
DATA   indexed in order of last differences
'          --index inverse

    DEFINT A-Z: CLS
    READ a$, b$: PRINT a$: PRINT b$
In:
    INPUT "m = "; m: IF m = 0 THEN END
    INPUT "index = "; i
    PRINT "the combination is"
    DIM p(m): FOR j = 1 TO m: p(j) = j: NEXT
    i = i - 1: h = m: GOSUB Inv: CLEAR : GOTO In

Inv:
    IF i = 0 THEN GOSUB Fin: RETURN
    r = 0: f = 1
O:
    r = r + 1: g = f * (m + r) \ r
    IF i >= g THEN f = g: GOTO O
    p(m) = m + r: i = i - f: m = m - 1: GOTO Inv

Fin:
    FOR j = 1 TO h: PRINT p(j); : NEXT: PRINT : RETURN
```

Combinations m at a time
indexed in order of last differences

m = ? 4
index = ? 11

the combination is
  1 3 5 6

Bibliography

Afriat, S. N. (1950). The quadratic form positive definite on a linear manifold. *Proc. Cambridge Phil. Soc.* 47, 1, 1-6.

(1954a). Composite matrices. *Quart. J. Math. Oxford* 5, 18, 81-98.

(1954b). Symmetric matrices, quadratic forms and linear constraints. *Publicationes Mathematicae* 3, 3-4, 305-308.

(1957a). On the definition of the determinant as a multilinear antisymmetric function. *Publicationes Mathematicae* 5, 38-39.

(1957b). Orthogonal and oblique projectors and the characteristics of pairs of vector spaces. *Proc. Cambridge Phil. Soc.* 53, 4, 800-816.

(1960). The system of inequalities $a_{rs} > x_r - x_s$. *Research Memorandum* No. 18 (October), Econometric Research Program, Princeton University. *Proc. Cambridge Phil. Soc.* 59, 1963, 125-133.

(1961). Gradient Configurations and quadratic functions. *Research Memorandum* No. 20 (January), Econometric Research Program, Princeton University. *Proc. Cambridge Phil. Soc.* 59, 1963, 287-305.

(1964). The construction of utility functions from expenditure data. *Discussion Paper* No. 144 (October 1964), Cowles Foundation, Yale University; First World Congress of the Econometric Society, Rome, September 1964; *International Economic Review* 8, 1 (1967), 67-77.

(1968). The construction and characteristic properties of the general matrix inverse (mimeo). University of North Carolina, Chapel Hill.

(1971). Theory of maxima and the method of Lagrange. SIAM *J. Appl. Math.* 20, 3, 343-357.

(1973). On the rational canonical form of a matrix. *Linear and Multilinear Algebra* 1, 185-6.

(1974). Sum-symmetric matrices. *Linear Algebra and Its Applications* 8, 129-40.

(1975). The Algebra and Geometry of Statistical Correlation. In Afriat, Sastry and Tintner (1975), Part I, 1-100.

(1980). *Demand Functions and the Slutsky Matrix.* Princeton University Press. (Princeton Studies in Mathematical Economics, 7)

(1981). On the constructability of consistent price indices between several periods simultaneously. In *Essays in Theory and Measurement of Demand: in honour of Sir Richard Stone*, edited by Angus Deaton. Cambridge University Press. 133-61.

(1987). *Logic of Choice and Economic Theory.* Oxford: Clarendon Press.

(1988). Lagrange Multipliers. In *The New Palgrave: a Dictionary of Economic Theory and Doctrine*, edited by John Eatwell, Murray Milgate and Peter Newman. Macmillan.

— , M. V. Rama Sastry and Gerhard Tintner (1975). *Studies in Correlation: Multivariate Analysis and Econometrics.* Göttingen: Vandenhoeck and Ruprecht.

Aitken, A. C. (1932). Turnbull and Aitken (1932).
  (1942). *Determinants and Matrices.* Edinburgh and London: Oliver and Boyd.

Althoen, Steven C. and Renate McLoughlin (1987). Gauss-Jordan Elimination: a Brief History. *Amer. Math. Monthly 94* (February), 130-42.

Bell, E. T. (1940). *The Development of Mathematics.* McGraw-Hill. Ch. 9, 186-92; 2nd ed., 1945, 203-9.

Bellman, Richard (1960). *Introduction to Matrix Analysis.* New York: McGraw-Hill.

Bodewig, E. (1956). *Matrix Calculus.* North-Holland. 105-9; 2nd ed., 1959, 121-5.

Boyer, Carl B. (1968). *A History of Mathematics.* Wiley. 218-9.

Cottle, R. W. and C. E. Lemke, eds. (1976). Nonlinear Programming. SIAM-AMS *Proc.* IX.

Dantzig, George B. (1963). *Linear Programming and Extensions.* Princeton University Press.

De Finetti, Bruno (1949). Sulle stratificazioni convesse. *Ann. Matematica Pura App.* 4, 173-83.

Egerváry, Eugen (1960). *Z. Agnew Math. Phys.* 11, 376-86.

Fenchel, W. (1953). *Convex cones, sets and functions.* Notes by D. W. Blackett of lectures delivered in the Mathematics Department, Princeton University.

Ferrar, W. L. (1951). *Finite Matrices.* Oxford: Clarendon Press.

Finsler, P. (1937). Über das Vorkommen definiter und semidefiniter Formen in schären quadratischen Formen. *Commentarii Mathematicii Helveticii* 9, 188-92.

Ford, L. R. Jr and D. R. Fulkerson (1962). *Flows in Networks.* Princeton University Press.

Forsythe, G. E. and C. B. Moler (1967). *Computer Solution of Linear Algebraic Systems.* Prentice Hall. 27-36.

Fox, L. (1965). *An Introduction to Numerical Linear Algebra.* Oxford University Press. 60-5, 68-73, 87-91, 99-109.

Gale, David (1960). *The Theory of Linear Economic Models.* McGraw-Hill.

Gantmakher, F. P. (1959). *Applications of the Theory of Matrices.* New York and London: Interscience. Translation by J. L. Brenner of *Teoriya Matrits;* Gostekhizdat, 1953; 2nd part.

Gewirtz, Allan, Harry Sitomer and Albert W. Tucker (1974). *Constructive Linear Algebra.* Englewood Cliffs NJ: Prentice Hall.

Gibbs, J. Willard (1886). *Proc. A.A.A.S.* 35, 37-66.
(1891). *Nature* 44, 79-82.
(1906). *Scientific Papers II.* Longmans, Green & Co. Pp. 91-117; 161-81; 53-4, 112, 166-7.

Golub, Gene H. and Charles F. Van Loan (1983). *Matrix Computations.* Baltimore: Johns Hopkins Press.

Grassmann (1844). *Ausdehnungslehre* (1st).
(1862). *Ausdehnungslehre* (2nd).

Grimshaw, M. E. (1951). Hamburger and Grimshaw (1951).

Halmos, Paul R. (1948). *Finite Dimensional Linear Space.* Princeton University Press. 2nd Edition, Van Nostrand, 1958.

Hamburger, H. L. and M. E. Grimshaw (1951). *Linear Transformations in n-dimensional Vector Space.* Cambridge University Press.

Householder, Alston (1964). *Theory of Matrices in Numerical Analysis.* New York: Blaisdell Publ. Co. (Dover reprint, 1975). Pp. 5-6, 13, 24, 125-7, 142.

Jacobson, N. (1953). *Lectures in Abstract Algebra,* Vol. II: Linear Algebra. New York: Van Nostrand. Pp. 8, 9, 13.

Kemeny, J. G. (1946). *See* Kirchhoff (1847).

Kemeny, J. G., J. L. Snell and G. L. Thompson (1974). *Introduction to Finite Mathematics,* 3rd edition. Prentice-Hall. 336-359.

Kirchhoff, G. R. (1847). Solution of the equations to which one is led through the investigation of electric currents in a linear circuit. *Poggendorf Annalen* v, 72 (tranlation by J. G. Kemeny, mimeo, 1946).

Kuhn, H. W. (1976). Nonlinear programming: a historical view. In Cottle and Lemke (1976), 1-26.
(1983). Complementarity problems—an eclectic view. Nordic LCP Symposium *Proceedings,* Linkping University, Sweden, March 23-7.

— and Albert W. Tucker (1956). Linear Inequalities and Related Systems. *Annals of Mathematics Studies* No. 38, Princeton University Press.

McDuffee, C. C. (1943). *Vectors and Matrices.* Buffalo, NY: The Mathematical Association of America.
(1946). *The Theory of Matrices.* New York: Chelsea.

Marcus, Marvin and Henryk Minc (1964). *A Survey of Matrix Theory and Matrix Inequalities.* Boston: Allyn and Bacon.
(1965). *Introduction to Linear Algebra.* New York and London: Macmillan.

Mirsky, L. (1955). *An Introduction to Linear Algebra.* Oxford: Clarendon Press.

Muir, Thomas (1906). *History of Determinants.* London

Noble, B. (1969). *Applied Linear Algebra.* Prentice Hall. 211-22.

Polya, George. *How to solve it: a new aspect of mathematical method.* Princeton University Press, 1973.

Schreier O. and E. Sperner (1955). *Introduction to Modern Algebra and Matrix Theory.* New York: Chelsea.

Sitomer, Harry (1974). Gewirtz, Sitomer and Tucker (1974).

Strang, Gilbert (1976). *Linear Algebra and Its Applications.* Academic Press. pp. 91, 137; 2nd ed. 1980, pp. 97, 144.

Stiefel, Eduard L. (1963). *An Introduction to Numerical Mathematics.* New York: Academic Press.

Stewart, G. W. (1973). *Introduction to Matrix Computations.* New York: Academic Press. 113-44.

Sylvester, J. J., *Nature* 31, p. 35.

Todd, J. A. (1947). *Projective and Analytical Geometry.* London: Pitman.

Tucker, Albert W. (1947). Notes on the Algebraic Topology of Kirchhoff's Laws (mimeo). February.
   (1951). On Kirchhoff's Laws, Potential, Lagrange Multipliers, Etc. Mathematics Department, Princeton University, and National Bureau of Standards, Los Angeles, Institute of Numerical Analysis, August 6.
   (1960a). A Combinatorial Equivalence of Matrices. Proceedings of Symposia in Applied Mathematics, Volume X, *Combinatorial Analysis.* American Mathematical Society. pp. 129-40.
   (1960b). Combinatorial Problems. *IBM Journal of Research and Development* 4, 5 (November), 454-560.
   (1974). Gewirtz, Sitomer and Tucker (1974).
   (1975a). A View of Linear Programming Theory. Meeting of Japanese O.R. Society, Tokyo, August.
   (1975b). CR-decomposition of a matrix (mimeo). Mathematics Department, Princeton University, December.
   (1976a). Combinatorial Mathematics (mimeo). May.
   (1976b). Least squares extension of linear programming. *Proceedings of the 9th International Mathematical Programming Symposium, Budapest, August 23-7, 1976,* edited by A. Prekopa. North-Holland, 1979.
   (1978). On Matrix Decomposition—Computation, Duality, History, Theory. August 4.
   (1979a). On Matrix Multiplication (mimeo). January 10.
   (1979b) (with Evar D. Nering, Arizona State University). Constructive proof of a general duality theorem for linear inequalities and equations. 10th International Symposium on Mathematical Programming, Montreal, August 27.

(1979c). On matrix multipliers (mimeo). Department of Mathematics, Princeton University, October 1.

(1983a). Constructive LP theory—with extensions. Nordic LCP Symposium *Proceedings*, Linköping University, Sweden, March 23-27. 169-72.

(1983b). Decomposition of a matrix into rank-determining matrices. *College Mathematics Journal* 14, 3 (June), 232.

(1984a). Matrix decomposition for echelon bases (mimeo). Stanford University, May.

(1984b). Dual basic factors of a matrix. ICME 5 *Abstracts*, Adelaide, South Australia, August.

(1986). On matrix decomposition (mimeo). Rutgers Centre for Operations Research, NJ. May 19.

(1987a). Tableau semi-inversion and decomposition (mimeo). Princeton, March.

(1987b). Private communication, Princeton, 14 December.

(1988). Another Look at Matrix Decomposition. February 8.

Turnbull, H. W. (1945). *The Theory of Determinants, Matrices and Invariants*, 2nd edition. London and Glasgow: Blackie.

— and A. C. Aitken (1932). *An Introduction to the Theory of Canonical Matrices*. London and Glasgow: Blackie.

Weierstrass. *Werke* III, 271-86.

Weyl, Hermann (1939). *The Classical Groups*. Princeton University Press.
(1946). Princeton University Bicentennial Conference, December 17-19.

Willner, L. B. (1967). *Math. Comp.* 21, 227-9.

Zurmühl, Rudolph (1949). *Matrizen*.

Index